WHAT THIS **Super** WILL DO FOR YOU

This **Super Review** provides all that you need to know to do your homework effectively and succeed on exams and quizzes.

The book focuses on the core aspects of the subject, and helps you to grasp the important elements quickly and easily.

Outstanding **Super Review** features:

- Topics are covered in logical sequence

- Topics are reviewed in a concise and comprehensive manner

- The material is presented in student-friendly language that makes it easy to follow and understand

- Individual topics can be easily located

- Provides excellent preparation for midterms, finals and in-between quizzes

- In every chapter, reviews of individual topics are accompanied by Questions **Q** and Answers **A** that show how to work out specific problems

- At the end of most chapters, quizzes with answers are included to enable you to practice and test yourself to pinpoint your strengths and weaknesses

- Written by professionals and test experts who function as your very own tutors

Dr. Max Fogiel
Program Director

CONTENTS

REA's Books Are The Best...
They have rescued lots of grades and more!

(continued from previous page)

"Your books have saved my GPA, and quite possibly my sanity. My course grade is now an 'A', and I couldn't be happier."

Student, Winchester, IN

"These books are the best review books on the market. They are fantastic!"

Student, New Orleans, LA

"Your book was responsible for my success on the exam. . . I will look for REA the next time I need help."

Student, Chesterfield, MO

"I think it is the greatest study guide I have ever used!"

Student, Anchorage, AK

"I encourage others to buy REA because of their superiority. Please continue to produce the best quality books on the market."

Student, San Jose, CA

"Just a short note to say thanks for the great support your book gave me in helping me pass the test . . . I'm on my way to a B.S. degree because of you!"

Student, Orlando, FL

Super Review®

All You Need to Know!

MICROBIOLOGY

By the Staff of
Research & Education Association
Dr. M. Fogiel, Director

Research & Education Association
61 Ethel Road West
Piscataway, New Jersey 08854

SUPER REVIEW®
OF MICROBIOLOGY

Year 2004 Printing

Printed in the United States of America

Library of Congress Control Number 00-131294

International Standard Book Number 0-87891-190-1

I-4

10 ROLE OF MICROBES IN DISEASE

11   MICROBES IN THE ENVIRONMENT

12   MICROBES IN INDUSTRY

CHAPTER 1

History and Scope of Microbiology

1.1 Early History

Microbiology is the study of microbes (microorganisms), i.e., organisms too small to be observed by the naked eye, and dates back to the seventeenth century, when Hans and Zaccharias **Janssen** (ca. 1600) invented the first compound microscope.

Robert **Hooke** (1665) made early observations using a compound microscope. From his observations of cork, he coined the word "cell" to describe the "little boxes" he saw as the smallest structures of life, setting the foundation for "cell theory."

The **cell theory** states that all living things are composed of cells.

Anton **van Leeuwenhoek** (1670s and 1680s) was first to observe and describe living microbes, which he referred to as "animalcules." His homemade microscopes magnified specimens up to 200x–300x.

Carolus **Linnaeus** (1735) developed a general system of classification and **binomial nomenclature** (genus name + specific epithet or species name).

The theory of **spontaneous generation**—which states that new life can arise from nonliving matter—was commonly accepted until the mid-1880s.

Francesco **Redi** (1660s) was first to present experimental evidence to refute spontaneous generation. Redi used cloth-covered jars to show that maggots do not arise spontaneously in meat but that the meat must be open to contact with flies in order for maggots to appear. However, his results were not accepted by many (the experiment that finally put the theory of spontaneous generation to rest did not occur for another 200 years!—see **Pasteur**, section 1.2).

John **Needham** (1740s) published experimental evidence in support of spontaneous generation. In his experiments, he showed that nutrient solutions could be boiled, and yet when cooled, microorganisms would soon appear. In the 1760s, Lazzaro **Spallanzani** countered this with experiments demonstrating that if such flasks were sealed, the microorganisms did not appear after boiling. **Needham** and his supporters argued that the boiling was responsible for killing some "vital force" (later thought to be oxygen upon its discovery) and that sealing the flask prevented its re-entry. It was still another hundred years before Pasteur's experiments disproved the theory.

On the medical front, Edward **Jenner** (1798) developed and tested the first **vaccine**. Jenner noticed that milkmaids exposed to cowpox rarely developed the more serious smallpox. He then used cowpox to successfully inoculate patients against smallpox.

Ignaz Philipp **Semmelweis** (1840s) noticed a connection between doctors doing autopsies and patients developing **puerperal (childbirth) fever**; he was first to suggest that doctors should wash their hands between procedures.

1.2 The Golden Age of Microbiology

The **Golden Age of Microbiology** (1850-1890) was a period during which major historical figures established microbiology as a viable scientific discipline.

Louis **Pasteur**—disproof of spontaneous generation/proof of biogenesis (1861). Pasteur devised a special kind of flask (the **swan necked flask**) in order to disprove spontaneous generation. Swan neck flasks are not sealed off; rather, the neck of the flask is open, but is long and curved. Such flasks are open to air and to any "vital force." However, microorganisms from the outer air become trapped in the curved neck of the flask and are thus prevented from contaminating the medium. Infusions or nutrient broths that have been sterilized by boiling do not become contaminated in such a flask unless the neck becomes broken. Thus, Pasteur disproved spontaneous generation while demonstrating that the inoculating (contaminating) organisms are present in the air.

Pasteur was first to show that microorganisms are *everywhere*, including the air. This discovery provided impetus for the development of **aseptic techniques** in the laboratory and medical situations to prevent contamination.

Pasteur was also instrumental in work on the role of yeast and other microorganisms in **fermentation** or the conversion of sugars to alcohol. He developed a heating process used to kill bacteria in some alcoholic beverages and milk, i.e., **pasteurization**.

Pasteur was also a pioneer in the area of immunology. He developed "vaccines" (he coined the term) for chicken cholera and rabies (1884). In his search for a rabies vaccine, Pasteur used the brain tissue of rabid animals to inoculate rabbits. He then used the dried spinal cords of those rabbits to inoculate experimental animals. In 1865, he used this treatment to successfully vaccinate a young boy who had been bitten by a rabid dog, and showing signs of the disease, was expected to die. The vaccine worked and the boy lived. (Modern **vaccines** are generally live, avirulent microorganisms, or killed pathogens or components isolated from pathogens, especially by use of recombinant DNA techniques.)

It was one of Pasteur's publications (1857) that laid the foundation for the **germ theory of disease** by suggesting that microorganisms are the *cause* of disease rather than the *result* of it. This theory states that microorganisms can invade other organisms and are responsible for the transmission of infectious diseases.

Joseph **Lister** (1860s) introduced the use of disinfectants to clean surgical dressings and instruments. Robert **Koch's** work (1870s) provided further support for the germ theory of disease. His work with the sheep disease **anthrax** was instrumental in establishing the concept of "one disease—one organism," which is the foundation of medical microbiology. He was the first to establish **pure culture technique**, and the first to use agar in growth medium. **Koch's postulates** (1876; see section 10.5) are still used today as the appropriate method for demonstrating that a specific microorganism transmits a specific disease.

1.3 Later Discoveries and the Beginnings of Virology

Virology had its beginnings in 1892 when Dmitri **Ivanowski** showed that the organism responsible for tobacco mosaic disease was able to pass through filters that stopped all known bacteria. Wendell **Stanley** (1935) later demonstrated that this organism was fundamentally different from other known organisms and could even be crystallized. The development of the electron microscope in the 1940s paved the way for further understanding and study of viruses.

Paul **Ehrlich** (1910) was first to use chemicals to treat disease (chemotherapy). His search for a "magic bullet" that would kill the cells of the infecting microorganism but not those of the host led to the discovery of Salvarsan in 1910. This chemotherapeutic agent was used in the treatment of syphilis. Along with Ehrlich's work, the discoveries of penicillin in 1928 by Alexander **Fleming** and sulfa drugs in 1932 by Gerhard **Domagk** ushered in our modern era of chemotherapy.

1.4 Scope of Microbiology

Microorganisms are found in all five kingdoms: Monera, Protista, Fungi, Plantae, and Animalia. All of the Monera (bacteria and cyanophytes) are microorganisms.

There are many fields of microbiology. Study of the classification of microorganisms constitutes **microbial taxonomy**. **Bacteriology** is the study of bacteria; **phycology**, the study of algae; **mycology**, the study of fungi; **protozoology**, the study of protozoans; **parasitology**, the study of parasitic microbes; and **virology**, the study of viruses.

Epidemiology is the study of the distribution and frequency of disease. **Etiology** is the study of the causes of disease. **Immunology** is the study of host defense against disease.

Various fields of **applied microbiology** include food technology, industrial microbiology, medical/pharmaceutical microbiology, genetic engineering, and environmental microbiology.

Microbial metabolism, microbial genetics, and **microbial ecology** are all fields of microbiology as well.

CHAPTER 2

Equipment and Techniques

2.1 Units of Measurement

The basic unit of measure is the **meter (1 m = 3.28 ft)**. There are 1,000 millimeters (mm) per meter.

Microbes are generally measured in micrometers (μm) or nanometers (nm).

$$1 \ \mu m = 0.000001 = 10^{-6} \ m$$
$$1 \ nm = 10^{-9} \ m$$

2.2 Microscopes

Microscopy is the process of projecting energy (visible light, ultraviolet light, or electrons) onto an object, and then using the energy that is emitted from that object to create an image on a sensing device (e.g., the lens of your eye, a screen, or a photographic film). Microscopes may differ in the kind of energy used, type of sensing device, resolution, wavelength, and magnification.

Resolution refers to the ability to distinguish adjacent objects or structures as separate and discrete entities. The **resolving power** of a microscope indicates the size of the smallest objects that can be clearly observed.

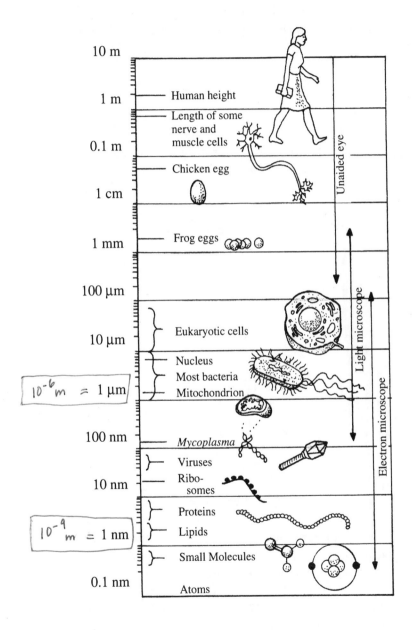

10 m

1 m — Human height

— Length of some
0.1 m nerve and
muscle cells

— Chicken egg

1 cm

1 mm — Frog eggs

100 μm

10 μm Eukaryotic cells

Nucleus
Most bacteria
Mitochondrion

10^{-6} m ≈ 1 μm

100 nm — *Mycoplasma*

Viruses
Ribo-
10 nm somes

Proteins

10^{-9} m ≈ 1 nm Lipids

Small Molecules

0.1 nm
Atoms

Unaided eye

Light microscope

Electron microscope

Figure 2.1 Relative Sizes

The limiting factors in resolution are the **wavelength** (λ) of the energy source (e.g., the length of light rays) and the numerical aperture (N.A.) of the lenses.

$$\text{resolving power} = \frac{\lambda}{2 \text{ N.A.}}$$

A **compound light microscope** uses a two-lens system—an ocular lens and an objective lens. Total **magnification** (magnifying power) equals the magnification of the ocular lens *multiplied by* that of the objective lens.

In **bright-field microscopy,** the light is transmitted directly through the specimen. Organisms must generally be stained.

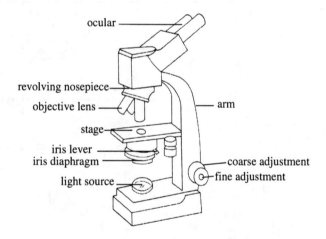

Figure 2.2 The Compound Microscope

Dark-field microscopes have a condenser that reflects light off the specimen at diverse angles and prevents most light from passing directly through the specimen. At the apex of the illuminating cone, light rays are scattered from the specimen into the objective lens and the image is that of a light organism against a dark background. This type of microscope is good for visualizing capsules and in diagnosing diseases caused by spiral bacteria. Even though these organisms are near the limit of resolution, their characteristic movement is discernible in dark-field.

The **phase-contrast** microscope is preferred for the observation of living, unstained organisms. A condenser splits the light beam, throwing the light rays slightly out of phase. Small differences in the densities and refractive indexes of various structures are accentuated, and internal details of live, unstained cells can be observed.

DIC

Nomarski (differential interference contrast) microscopes are much like phase-contrast, but the depth of field is very short, resulting in much greater resolution. Images appear to be nearly three-dimensional.

In **fluorescence microscopy,** ultraviolet light is used to excite electrons in molecules. When these electrons fall back to their original energy state, they fluoresce (emit light). Some organisms are naturally fluorescent, but most organisms observed in this way must first be treated with a fluorescent dye or **fluorochrome** (e.g., fluorescein).

A special technique known as **fluorescent antibody staining** or **immunofluorescence** is sometimes used to diagnose an unknown organism. In this case, the fluorochrome is attached to an antibody. Fluorescence indicates that the corresponding antigen is present, thus yielding a positive identification.

Electron microscopes use energy in the form of a beam of electrons rather than a beam of light, and the beam is focused using electromagnets rather than lenses. There are two types of electron microscopes: the transmission electron microscope and the scanning electron microscope. Viruses can be seen only by using electron microscopes.

In **transmission electron microscopy (TEM),** electrons pass through the specimen. Consequently, TEM cannot be used to view whole organisms; rather, very thin slices (ultrathin sections, e.g., 0.07 μm) are used. TEM sections are generally coated with a heavy metal. TEM is very good for looking at internal structures, and can resolve objects as close as 1 nm, and magnify up to 100,000 times. By using additional photographic enlargement, a total magnification of up to 20 million times may be achieved.

Scanning electron microscopy (SEM) is used to create three–dimensional images of surfaces, both external (whole organism) and internal (via freeze-fracture techniques—see below). SEM specimens are usually coated with an Au/Pd (gold/palladium) alloy or carbon. A magnification of up to 10,000 times, with a resolution of 20 nm, can be achieved (prior to further photographic enlargement). In SEM, electrons are released from the surface of the specimen rather than passing through it. Thus, SEM can provide topographical views of the surfaces of objects in their natural state (without sectioning).

Freeze-fracturing refers to a process wherein a cell is frozen and then cleaved with a knife. This technique can be used with either TEM or SEM to view surfaces or internal structures.

Freeze-etching is a further modification wherein water is evaporated from a freeze-fractured specimen to expose additional surfaces.

Table 2.1 Comparison of Microscopes

	Type of energy	Wavelength	Resolution	Maximum magnification
Light microscope	Visible light	400–700 nm	220 nm	usually 100×, (1,000×–2,000× with oil immersion)
Fluorescence microscope	Ultraviolet light	100–400 nm	110 nm	as above
Electron microscope	Electrons	0.005 nm	1 nm (TEM) 20 nm (SEM)	100,000× (TEM) 10,000× (SEM) prior to photographic enlargement

2.3 Preparation of Specimens for Light Microscopy

Wet mount—a drop of medium containing the organisms is placed on a microscope slide. This technique is used to view living organisms. Carboxymethyl cellulose (2%) can be added to the drop to slow the movement of fast-moving microorganisms.

Hanging drop—a drop is placed on a cover slip that is then inverted over a depression slide. This technique is sometimes used for dark-field illumination.

Smear—used to view dead organisms (the organisms are killed by the process). Microorganisms from a drop of medium are spread across the surface of a glass slide. This smear is air-dried and then heat-fixed (by passing through flame). **Heat fixation** accomplishes three things: (1) it kills the organisms, (2) it causes the organisms to adhere to the slide, and (3) it alters the organisms so that they more readily take up stains.

2.4 Staining

Stain (dye)—a molecule that can bind to a cellular structure and give it color, distinguishing it from the background. In addition to discerning parts of the cell, staining helps to categorize microorganisms, and is used to examine structural and chemical differences in their cell walls. Stains are used extensively in bacterial identification and classification.

Basic (cationic) dyes—positively charged dyes. Examples include methylene blue, crystal violet, safranin, and malachite green. Most bacteria have negatively charged surfaces to which these positively charged dyes are attracted.

Acidic (anionic) dyes—negatively charged dyes (e.g., eosin, picric acid, nigrosin, Congo red). These dyes are often used in **negative staining** wherein the background is stained rather than the organism. With this technique, the cells appear clear against a colored background. It is often used to view capsules. Negative staining avoids heat fixation and chemical reactions; thus, cells appear more natural and less distorted.

Stains may be either **simple** or **differential**.

Simple stain—a single dye is used. It reveals basic cell shapes and arrangements. Examples include methylene blue, safranin, carbolfuchsin, and gentian violet.

Differential stain—two or more dyes are used to distinguish between two kinds of organisms or between two different parts of the same organism. Examples include the Gram stain, Schaeffer-Fulton spore stain, and Ziehl-Neelsen acid-fast stain.

The Gram stain (Christian Gram, 1884)—reveals fundamental differences in the nature of the cell wall (probably due to differences in the amount of peptidoglycan in the cell wall—gram-positive bacteria have more peptidoglycan than do similar gram-negative ones). Often used in bacterial taxonomy, the Gram stain is used on air-dried, heat-fixed (therefore killed) specimens.

2.4.1 The Gram Stain Procedure

The Gram stain procedure includes the following steps (in order):

1. Cells are stained with crystal violet followed by iodine— *all* cells appear purple.

2. Cells are washed with alcohol/acetone—gram-positive organisms remain purple, gram-negative organisms are now clear.

3. Safranin is added to the cells—gram-positive organisms remain purple, gram-negative organisms now appear pink or orange-red.

2.4.2 Mordant

A **mordant,** often an iodine solution, may be added to a staining solution to intensify the stain. The mordant increases the stain's affinity for the specimen or coats the specimen to increase its size and staining ability.

2.4.3 Special Stains

Staining techniques may be specialized for the staining of the capsule, endospores, and flagella.

The **capsule** is a gelatinous coat and is often used as an indicator of virulence. The capsule is water soluble and therefore is removed in normal staining procedures. The capsule is visualized by a staining technique called **negative staining,** in which the specimen is first stained with India ink or nigrosin, which provides a dark background for contrast. The first stain is followed by a simple stain, which appears as a light region around each individual organism.

Endospores are dormant structures that are stained by malachite green and a heating technique to allow the dye to penetrate the cell and the endospores. The endospores are specifically stained by this procedure. Again, to add contrast, a second simple stain usually is added. Thus, endospores appear green within the cellular red/pink background.

Flagella are whiplike structures used in locomotion that require specialized staining methods to increase their thickness. A mordant such as tannic acid or potassium alum is used for this purpose followed by the stain pararosaniline (Leifon method) or carbolfuchsin (Gray method).

Problem Solving Examples:

 Explain why the bacterial cell wall is the basis for Gram staining.

Bacteria stained by this method fall into two groups: gram-positive, which retain the crystal violet-iodine complex and are dyed a deep violet color, and gram-negative, which lose the crystal violet-iodine complex when treated with alcohol and are stained by the safranin, giving them a red color.

The cell walls of bacteria are composed of peptidoglycan. This layer is thick in gram-positive cells, but thin in gram-negative cells.

However, the latter have a lipopolysaccharide layer surrounding the thin peptidoglycan layer.

The difference in staining is due to the high lipid content (about 20%) of the cell walls of gram-negative bacteria. (Lipids include fats, oils, steroids, and certain other large organic molecules.) During the staining procedure of gram-negative cells, the alcohol treatment extracts the lipid from the cell wall, resulting in increased permeability of the wall. The crystal violet-iodine complex is thus leached from the cell in the alcohol wash. The decolorized cell then takes up the red safranin. The cell walls of gram-positive bacteria have a lower lipid content, and thus become dehydrated during alcohol treatment. Dehydration causes decreased permeability of the wall, so that the crystal violet-iodine complex cannot leave the cell, and the cell is thus violet-colored. In addition, a thick peptidoglycan layer of the gram-positive cell walls prevents decolorization by alcohol. Most bacteria fall into one of these two staining groups, and Gram staining is an important means of bacterial classification.

 Explain how a mordant can be used in the Gram staining technique.

In using the Gram staining technique to separate and classify bacteria, a mordant in the form of iodine solution is applied to identify gram-positive bacteria.

In the Gram staining technique used for staining and separating bacteria into gram-negative and gram-positive bacteria, a mordant in the form of iodine solution is applied to identify gram-positive bacteria. The mordant binds the crystal violet dye to the cell of gram-positive bacteria and helps resist decolorization so that the cell is more strongly stained. Gram-negative bacteria, however, do not retain the dye and are colorless.

2.5 Culturing Microorganisms

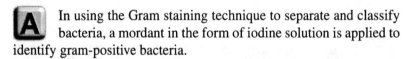

A **culture** is a population of microbes living in a culture medium.

Aseptic technique—the prevention of microbial contamination—is critical to pure culture of microorganisms.

2.5.1 Culture Media

A liquid or solid prepared for the purpose of growing microbes is called a nutrient broth or nutrient agar or **culture medium.** A culture medium generally provides the organisms with everything required for their growth, including a carbon source, a nitrogen source, and essential trace elements. Agar, a seaweed derivative, can be used to solidify the medium, which is then called nutrient agar.

A culture medium may be **defined**—in which known quantities of specific nutrients are used and the exact chemical composition is known, or **complex**—in which nutrients of reasonably well-known composition are used, but the exact composition varies from batch to batch, and thus, the exact chemical composition is not known. Synthetic media are chemically defined; natural media are usually complex.

[MSA] **Selective media** encourage the growth of some organisms while discouraging the growth of others. Salts, dyes, and antibiotics are among the substances used to inhibit growth of unwanted organisms. An example is mannitol salt agar, which encourages the growth of staphylococci while inhibiting the growth of most other bacteria.

[MAC] A **differential medium** is one that distinguishes growing microbes on the basis of visible chemical changes in the medium. Reagents in the medium allow for a specific chemical reaction to take place, but only in the presence of particular microorganisms. For example, MacConkey agar contains lactose and red dye; bacteria that can ferment the lactose take up the dye, resulting in the formation of red colonies. Bacteria that cannot ferment lactose form colorless colonies.

A medium can be both differential and selective.

Enrichment medium contains a nutrient (or nutrients) that enhances the growth of particular microbes, e.g., addition of blood to nutrient medium enhances growth of streptococci. This technique is designed to increase the relative numbers of a particular organism; it is possible to obtain pure cultures with this culture method.

Selective media, differential media, and enrichment media are all **diagnostic** media.

Reducing medium is used for the growth of anaerobic organisms. This type of medium contains reducing agents such as thioglycollate or cysteine that will deplete the oxygen levels in the medium.

Liquid medium is used in tightly capped test tubes and heated before use.

Solid medium must be contained in specialized, oxygen-free vessels such as **anaerobic glove boxes** or **anaerobic chambers** such as a GasPak jar, which uses hydrogen and a palladium catalyst to remove O_2 through the formation of water. Oxygen may also be removed using a vacuum pump and flushing out residual O_2 with nitrogen gas.

Problem Solving Example:

Design an experiment that would select for a nutritional mutant.

Among the large variety of bacterial mutants are those that exhibit an increased tolerance to inhibitory agents such as antibiotics. There are also mutants that exhibit an altered ability to produce some end product, and mutants that are nutritionally deficient (unable to synthesize or utilize a particular nutrient). These nutritional mutants are called auxotrophs because they require some nutrient not required by the original cell type or prototroph.

The first step in isolating a nutritional mutant is to increase the spontaneous mutation rate of the bacteria, which usually ranges from 10^{-6} to 10^{-10} bacterium per generation. (This means that only 1 bacterium in 1 million to 1 in 10 billion is likely to undergo a mutational change.) The mutation rate can be significantly increased by exposing a bacterial culture (e.g., *Escherichia coli*) to ultraviolet radiation or x-rays.

A portion of the irradiated *E. coli* culture is placed on the surface of a Petri dish and spread over the surface to ensure the isolation of colonies. This Petri dish must contain a "complete" medium, such as nutrient agar, so that all bacteria will grow, including the nutritional mutants. After incubation of this "complete" medium plate, the exact position of the colonies on the plate is noted. A replica plating device is then gently pressed to the surface of the complete plate, raised, and then pressed to the surface of a "minimal" media plate. The replica plating device transfers the exact pattern of colonies from plate to plate. The positioning of the cloth on the "minimal" agar plate must be identical to its positioning when originally pressed to the complete plate. Colony locations will then be comparable on each of the two plates, which are termed "replicas." The "minimal" medium consists only of glucose and inorganic salts, which are nutrients that normally permit the growth of *E. coli*. From these basic nutrients, normal *E. coli* can synthesize all required amino acids, vitamins, and other essential components.

After incubation, colonies appear on the minimal plate at most of the positions corresponding to those on the complete plate. Those missing colonies on the minimal plate are assumed to be nutritional mutants, because they cannot grow on a glucose and inorganic salts medium. These missing colonies can be located on the complete media plate by comparing the location of colonies on the replicas. If they had not been irradiated, and if mutations did not occur, all the colonies would have been able to grow on this minimal medium. The colonies that did grow were not affected by the irradiation and were nonmutants.

To determine the exact nutritional deficiency of the auxotrophs, one can plate them on media that contain specific vitamins or amino acids in addition to glucose and inorganic salts. These compounds are normally synthesized by prototrophic *E. coli*. The mutation might have affected a gene that controls the formation of an enzyme that, in turn, regulates a step in the biosynthesis of one of these nutrients. By plating the mutant on several plates with each plate containing an additional specific nutrient, we could determine which nutrient the mutant cannot synthesize. The mutant will grow on that plate containing the specific nutrient that the mutant is unable to synthesize.

2.5.2 Plate Methods

The **streak plate** and the **pour plate** are two techniques for obtaining pure cultures. In both cases, it is assumed that a single bacterial colony arises from a single cell. The number of cells is diluted by the medium in the pour plate method; in the streak plate technique, the number of cells is decreased manually by spreading the inoculum over the surface of the agar medium.

Inoculating loops are sterilized by passing them directly through a flame, to the point of redness. **Flaming** is also used to sterilize the tips of test tubes to insure that they are not contaminated during the inoculating procedure.

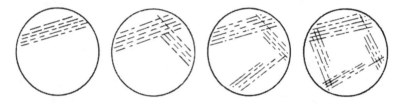

Figure 2.3 The Streak Plate. Microbes are quickly brushed or streaked across the cell-culture medium so as to allow cell growth on the plate surface.

Replica plating—velvet disks are used to pick up some of the cells from colonies on an agar medium. The disk is used to imprint the cells onto other agar plates that either lack nutrients or contain substances that select against (i.e., kill or stop the growth of) mutant cells. Colonies that do not grow on the replica plates are used to identify which of the colonies on the original plate contain the mutant cells. This technique is also used to identify clones.

Some organisms, including many parasites, cannot be cultured in synthetic media, but must be cultured in living cells. Sometimes this can be accomplished with cell or tissue culture techniques; in other cases, the organism must be cultured in a living animal.

Culture methods for molds and yeasts are similar to those used for bacteria.

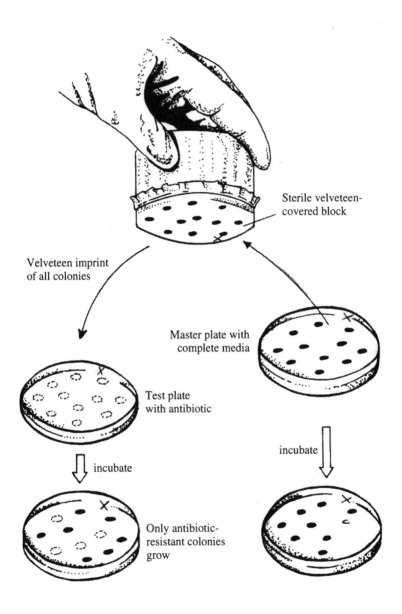

Sterile velveteen-covered block

Velveteen imprint of all colonies

Master plate with complete media

Test plate with antibiotic

incubate

incubate

Only antibiotic-resistant colonies grow

Figure 2.4 Replica Plating

2.5.3   Serial Dilutions of Broths

Proper dilution of cultures is important for accurate enumeration of microorganisms. One milliliter (ml) of a bacterial suspension is added to 9 ml sterile water, resulting in a 1:10 dilution of the original sample. One milliliter of the 1:10 dilution is added to 9 ml sterile water, yielding a 1:100 dilution. The number of bacteria are reduced one-tenth each time. The process can be repeated to achieve 1:1,000, 1:10,000, 1:100,000 dilutions, etc. The number of cells in 1 ml of the original sample is equal to the number of cells (or colonies) counted, multiplied by the reciprocal of the dilution.

2.5.4   Special Culture Techniques

Mycobacterium leprae, the leprosy bacillus, requires low temperatures for growth and has been cultured in armadillos.

Rickettsias and chlamydiae require a living host cell to complete their life cycles.

Carbon dioxide incubators are often used in laboratories to maintain specialized carbon dioxide levels for different organisms. For example, the organisms responsible for brucellosis, gonorrhea, and meningococcal meningitis require high carbon dioxide concentrations in culture.

2.5.5   Preserving Bacterial Cultures

Bacterial cultures may be stored for short or long terms. For short-term storage (days to weeks), refrigeration is the method used most often. For long-term storage (weeks to years), the rapid deep freezing of the specimen, usually with liquid nitrogen, is used. **Lyophilization** is also used, which involves freeze-drying the specimen such that all water is removed. The result of lyophilization is a powder material. Both freeze-drying and lyophilization allow for the storage of microbial cultures for years without loss of viability or an accumulation of mutations.

Problem Solving Examples:

Q Irradiation and incubation of an autotrophic leu-strain of bacteria until they reached the stationary phase were used to determine the spontaneous and x-ray induced mutation rate of leu⁻ and leu⁺. Also studied was the reaction of a control culture that was not irradiated. The cultures were then serially diluted, and 0.1 ml of various dilutions were plated on minimal medium plus leucine and on minimal medium.

Culture	Medium	Dilution	Number of Colonies
Irradiated	Minimal medium + leucine	10^{-9}	24
	Minimal medium	10^{-2}	12
Control	Minimal medium + leucine	10^{-9}	12
	Minimal medium	10^{-1}	3

What was the induced rate leading to prototrophic growth?

A The induced rate is the fraction value of the colonies formed on the minimal medium irradiated culture only divided by the colonies formed on the minimal medium plus leucine irradiated culture. The positive reciprocal of the dilution factor is used. The induced rate is therefore:

$$\frac{12 \times 10^2}{24 \times 10^9} = 5 \times 10^{-8}$$

Q Suppose you discovered a new species of bioluminescent worm. How could you prove that it was the worm itself and not some contaminating bacteria that was producing the light?

A   There are many bioluminescent organisms, but sometimes it is difficult to distinguish whether it is the actual organism that is emitting light or whether the light is due to luminous bacteria living in the organism. Some species of fish have light organs under their eyes where light-emitting bacteria live. One possible way of determining if a bacterium is the source of the light, is to take some of the light-producing substance and to place it in complete growth media (containing all possible nutrients). If the luminescent material proliferates, then a bacterium is most likely the causal factor. Another way would be to physically examine the light-emitting substance under a microscope. This could be done by scraping out some light-emitting substance from a bioluminescent organism and transferring it to a glass slide. If we see individual bacterium emitting light, then we can conclude that the bioluminescence is caused by bacteria.

Quiz: Equipment and Techniques

1. In preparing tissues for light microscopy, the sample must be fixed in order to

   (A) have a firm support.

   (B) distinguish subcellular regions from each other.

   (C) afford an unimpeded view of deep layers.

   (D) prevent autolysis.

   (E) provide a negative background in a counter stain.

2. Electron microscopes have higher resolution than light microscopes because

   (A) staining procedures are more efficient when the electron microscope is used.

   (B) electrons have a shorter wavelength than light.

   (C) the wave properties of light interfere with high resolution, whereas electrons have no wave properties.

(D)   the magnification of the electron microscope is much higher.

(E)   electrons have longer wavelengths than light.

3.   The Gram stain, which is used to differentiate bacterial cells, is based on

(A)   the protein content in the respective bacterial cell wall.

(B)   the carbohydrate content in the respective bacterial cell wall.

(C)   the lipid content in the respective bacterial cell wall.

(D)   the diffusion rate of staining fluid through the bacterial cell wall.

(E)   None of the above.

4.   Electron microscopy differs from light microscopy in the

(A)   staining technique used on the material.

(B)   preparation of material.

(C)   material studied.

(D)   resolution of the microscopes.

(E)   All of the above.

5.   Which of the following structures requires the use of an electron microscope for visualization?

(A)   A typical bacteria

(B)   The nuclear envelope and the internal membranes of mitochondria and chloroplasts

(C)   The 9 + 2 arrangement of microtubules typical of cilia and flagella

(D)   All of the above.

(E)   Both (B) and (C).

6. The primary advantage of phase-contrast and interference micros-
    copy is that they provide

    (A) particularly sharp resolution of surface contours.

    (B) resolution only slightly inferior to that of an electron
         transmission microscope, at a fraction of the cost.

    (C) the ability to observe living specimens.

    (D) a "full color" image.

    (E) the ability to detect intracellular regions of highly or-
         ganized structures by applying polarized light.

7. Which of the following techniques is/are used to prepare speci-
    mens for light microscopy?

    (A) Wet mount

    (B) Hanging drop

    (C) Smear

    (D) Heat fixation

    (E) All of the above.

8. The use of the light microscope in the study of fine cellular struc-
    ture is limited due to its

    (A) small size.

    (B) type of lenses.

    (C) difficulty of preparing materials.

    (D) lack of contrast.

    (E) relatively low power of resolution.

9. Which one of the following is the most stringent method for main-
    taining anaerobic microbes?

    (A) Use of reducing media

    (B) Use of an anaerobic container

(C)   Use of an anaerobic chamber

(D)   Filling bottles completely with media

(E)   Use of selective media

10.   Microbes are best preserved for long periods of time by

(A)   refrigeration.

(B)   lyophilization.

(C)   sublimation.

(D)   use of chemically defined media.

(E)   dehydration.

ANSWER KEY

1.	(A)	6.	(C)
2.	(B)	7.	(E)
3.	(C)	8.	(E)
4.	(E)	9.	(C)
5.	(E)	10.	(B)

CHAPTER 3

Survey of Microorganisms

3.1 Prokaryotes and Eukaryotes

Prokaryotes are organisms that lack a membrane-bound nucleus and membrane-bound organelles. All members of the Kingdom Monera are prokaryotes.

Eukaryotes are organisms in which the DNA (deoxyribonucleic acid) is enclosed within a membrane-bound nucleus; organelles are membrane-bound as well. All of the Protista, Fungi, Plantae, and Animalia are eukaryotes. Eukaryotic cells are usually larger and more complex than prokaryotic cells. Microscopic eukaryotes fall within the realm of microbiology. There are other characteristic differences between these two cell types (see Table 3.1); however, they are similar in chemical reactions and composition, e.g., both contain DNA and RNA (ribonucleic acid), plasma membranes, and ribosomes.

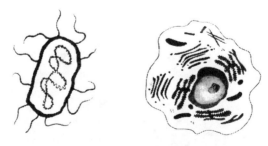

Figure 3.1 Typical Prokaryotic Cell, Left, and Eukaryotic Cell

Problem Solving Example:

 What advantages do multicellular organisms have over single-celled organisms?

Single-celled organisms represent one of the great success stories of evolution. They probably comprise more than half the total mass of living organisms and have successfully colonized even the harshest environments. Biochemically, many unicellular organisms, such as bacteria, are far more versatile than humans, being able to synthesize virtually everything they need from a few simple nutrients.

With the development of eukaryotic cells, certain evolutionary breakthroughs occurred. Eukaryotic cells are not only capable of being much larger than prokaryotes, they have the ability to aggregate into multicellular functional units. The cells of multicellular organisms are specialized for a variety of functions and interact in a way that make the organism more than the sum of its parts.

Multicellular organisms employ a high order of complexity. The constituent cells are specialized for the division of labor. The cells that make up a human body are not all alike; each is specialized to carry out certain functions. For example, red blood cells carry oxygen from the lungs to the various parts of the body while nerve cells are involved in the transmission of impulses. This specialization allows each cell to function more efficiently at its own task and also allows the organism as a whole to function more efficiently. Also, injury or death to a portion of the organism does not necessarily inhibit the functioning and survival of the individual as a whole.

Multicellular organisms are better adapted to survive in environments that are totally inaccessible to unicellular forms. This is most striking in the adaptation of multicellular organisms to land. Whereas unicellular animals and plants, such as the protozoans and the one-celled algae, survive primarily in a watery environment, the higher, multicellular organisms, such as the mammals and angiosperms, are predominantly land-dwellers. Multicellularity also carries the potential for diversity. Millions of different shapes, specialized organ systems,

Table 3.1 Comparison of Prokaryotes and Eukaryotes

	Prokaryotes	Eukaryotes
Membrane-bound organelles (including nucleus, mitochondria, chloroplast)	Absent	Present
Golgi apparatus	Absent	Present
Endoplasmic reticulum	Absent	Present
Chlorophylls, when present	Pigments in cytoplasm	Pigments located in membrane-bound chloroplast
Location of respiratory enzymes	Cell membrane	Mitochondria
Ribosomes	Small (70S*), free in cytoplasm	Larger (80S), bound to membranes within the cytoplasm
Plasma membrane	Fluid mosaic	Fluid mosaic with sterols
Cell wall, when present	Generally complex, contains peptidoglycan	Generally simple, may contain cellulose, chitin. Does not contain peptidoglycan
Flagellar protein	Flagellin	Tubulin
Cilia	Absent	Present
Mitosis	Lacking	Present
Genetic information	Single chromosome, without associated proteins, located in area called the nucleosome	Multiple chromosomes, with associated proteins, located in membrane-bound nucleus
Extrachromosomal DNA	In plasmids	In organelles such as the mitochondria and chloroplasts

*S is a sedimentation constant roughly proportional to MW.

and patterns of behavior are found in multicellular organisms. This diversity greatly increases the kinds of environments that organisms are able to exploit.

3.2 Bacteria

Bacteria are unicellular prokaryotes, ranging in diameter from about 0.20 to 2.0 µm. Bacteria reproduce by binary fission.

Until an official system of classification can be agreed upon, bacteriologists turn to *Bergey's Manual of Systematic Bacteriology* for practical classification of bacteria. *Bergey's Manual* was first published in 1923 as *Bergey's Manual of Determinative Bacteriology*. In *Bergey's Manual*, bacteria are identified and grouped on the basis of a number of characteristics, including: morphology (size, shape, arrangement), staining characteristics (gram-positive, gram-negative, acid-fast),

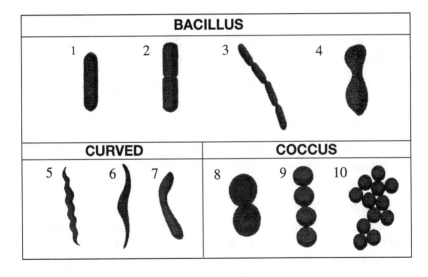

1	Single	5	Spirochete	8	Diplococcus
2	Diplobacillus (pair)	6	Spirillum	9	Streptococcus
3	Streptobacillus (chain)	7	Vibrio	10	Staphylococcus
4	Cocco-bacillus				

Figure 3.2 Primary Bacterial Shapes

nutritional requirements, growth characteristics, physical requirements (e.g., temperature and pH optima), biochemistry, genetics, spore-forming ability, type of movement, pathogenicity, antigen/antibody testing, DNA hybridization, and GC (guanine/cytosine) ratio.

The primary **bacterial shapes** are **bacillus** (rodlike), **coccus** (spherical), and **curved**. Bacilli in pairs are known as diplobacilli; in chains, they are called streptobacilli. Coccobacillus are bacilli shaped so short and wide that they resemble cocci. Pairs of cocci are diplococci; chains, streptococci. Irregular aggregations are staphylococci. Among the curved types are the **spirillum**, a flexible spiral, the **spirochete,** which forms a helix, and the comma-shaped **vibrio**. Bacteria that can assume several shapes are called **pleomorphic**.

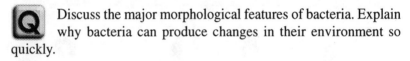

Problem Solving Examples:

Q Discuss the major morphological features of bacteria. Explain why bacteria can produce changes in their environment so quickly.

A Among the major characteristics of bacterial cells are their shape, arrangement, and size. These characteristics constitute the morphology of the bacterial cell.

Although there are thousands of different species of bacteria, the cells of most bacteria have one of three fundamental shapes: (1) spherical or ellipsoidal, (2) cylindrical or rodlike, and (3) spiral or helically coiled.

Spherical bacterial cells are called cocci (singular, coccus). Many of these bacteria form patterns of arrangement that can be used for identification. These patterns can be explained by peculiarities in the multiplication processes of the different bacteria. For example, diplococcal cells divide to form pairs. Streptococcal cells remain attached after dividing and form chains. Staphylococci divide three dimensionally to form irregular clusters of cocci, resembling bunches of grapes. Each of the cells in a diplococcal, streptococcal, or staphylococcal aggregate is an independent organism.

Cylindrical or rodlike bacterial cells are called bacilli (singular, bacillus). These do not form as wide a variety of arrangements as do cocci, but occasionally they are found in pairs or chains. These patterns do not arise from the multiplication process, but only from the particular stage of growth or growth conditions present, and hence bacilli usually appear as single, unattached cells. There are many variations in the thickness and length of these rodlike bacteria.

Spiral-shaped bacterial cells are called spirilla (singular, spirillum). Like the bacilli, they usually occur as unattached, individual cells. The spirilla exhibit considerable differences in length and in the frequency and amplitude of the spirals.

The average bacterial cell has dimensions of approximately 0.5 to 1.0 µm by 2.0 to 5.0 µm. (µm is the abbreviation for micrometer, which is $\frac{1}{1,000}$ of a millimeter or 10^{-6} meter.)

An important consequence of the very small size of a bacterial cell is that the ratio of surface area to volume is extremely high. This ratio allows a very large portion of the bacterial cell to be in contact with its environment. The result is that bacteria are able to rapidly ingest nutrients and growth factors and excrete wastes; their metabolic rate is correspondingly high. This high metabolic rate enables bacteria both to adjust to and to introduce changes in their environment in very short periods of time. These changes may be beneficial to the bacteria producing them or to other species of bacteria. For instance, the release of carbon dioxide by certain bacteria increases the acidity of the growth medium and favors the growth of bacteria requiring a low pH environment.

The most important factor involved in the ability of bacteria to alter their environment is their ability to multiply rapidly. Within the multitudes of bacteria produced by a newly settled bacterium, there will be a handful of mutants. Natural selection may allow one of these mutants to be most fit for survival in the changing environment, and numerous offspring will be derived from it. This cycle can be repeated, demonstrating the capability of the bacterial species to survive under changing conditions.

Is endospore formation in bacteria a method of reproduction? Explain.

Some bacteria have the capacity to transform themselves into highly resistant cells called endospores. In a process known as sporulation, bacteria form these intracellular spores in order to survive adverse conditions, such as extremely dry, hot, or cold environments. Each small endospore develops within a vegetative cell. The vegetative cell is the form in which these bacteria grow and reproduce. Each endospore contains DNA in addition to essential materials derived from the vegetative cell. All bacterial spores contain dipicolinic acid, a substance not found in the vegetative form. It is believed that a complex of calcium ions, dipicolinic acid, and peptidoglycan forms the cortex or outer layer of the endospore. This layer or coat helps the spore to resist the destructive effects of both physical and chemical agents. The dipicolinic acid/calcium complex may play a role in resuming metabolism during germination. The spore is generally oval or spherical in shape and smaller than the bacterial cell. Once the endospore is mature, the remainder of the vegetative cell may shrink and disintegrate. When the spores are transferred to an environment favorable for growth, they germinate and break out of the spore wall, and the germinating spore develops into a new vegetative cell. The endospore is incapable of growth or multiplication.

Endospore formation is neither a kind of reproduction nor a means of multiplication, since only one spore is formed per bacterial cell. During spore formation, a single endospore is present within the bacterium. The remaining portion of the vegetative cell dies off, while the endospore remains to later germinate into a new vegetative cell. This new vegetative cell is identical to the old one because it contains the same DNA. Spores only represent a dormant phase during the life of the bacterial cell. This phase is initiated by adverse environmental conditions.

Bacteria in the genera *Bacillus* and *Clostridium* are partially characterized by their ability to form endospores. Spores of *Bacillus anthracis,* the bacteria causing anthrax (primarily a disease of grazing animals), can germinate 30 years after they were formed.

3.2.1 Bacterial Anatomy

Structures **external** to the cell wall include the glycocalyx, flagella, and pili.

Glycocalyx—a polysaccharide-containing structure that functions in attachment to solid surfaces, preventing desiccation, and protection. It consists of a **capsule** and/or a **slime layer.**

Bacterial flagella—long, thin, filamentous structures composed of the protein **flagellin** that function in cell motility. Bacterial flagella originate in the cytoplasm at a basal body. Cells with one flagellum are monotrichous, those with many are multitrichous, and those lacking flagella are atrichous. Cells may differ in flagellar arrangement as well as in the number of flagella. [**Axial filaments (endoflagella)** are *subsurface* structures unique to spirochetes. They are similar to flagella, but wrap around the cell and cause the cell to move with a characteristic corkscrew-like motion.]

Flagella enable bacteria to move toward favorable conditions (**positive chemotaxis**) or away from unfavorable conditions (**negative chemotaxis**).

Pili are tiny hollow projections composed of a protein known as **pilin**. They are much smaller than flagella (visible only with electron microscopy or special techniques) and are not involved in cell motility. There are two kinds of pili: F pili, which function in the exchange of DNA during bacterial conjugation, and fimbriae, which are shorter than F pili and function in attaching the cell to other surfaces, enhancing colonization. Pili can affect infectivity—e.g., *Neisseria gonorrhoeae* with pili are highly infectious, but those lacking pili usually do not cause disease. Pili also serve as receptor sites for viruses.

The **cell wall** surrounds the plasma membrane and serves to protect the cell from changes in osmotic pressure, anchor flagella, maintain cell shape, and control transport of molecules into and out of the cell.

The typical **bacterial cell wall** is composed primarily of **peptidoglycan**, which is a polymer of two simple amino acid sugars: N-acetylglucosamine (NAG, gluNAc) and N-acetylmuramic acid (NAM, murNAc). Additional constituents include the **porins** and the **channel proteins**, which function in transport of molecules through the cell wall.

In addition to peptidoglycan, the cell walls of **gram-positive** bacteria also contain **teichoic acids**, polysaccharides that serve as attachment sites for bacteriophages (bacterial viruses).

Gram-negative cell walls are multilayered with a lipoprotein-lipopolysaccharide-phospholipid outer membrane external to the relatively thin peptidoglycan layer. This outer membrane protects the cell from antibiotics (e.g., penicillin) and enzymes (e.g., lysozyme).

Due to these differences in their cell walls, gram-negative and gram-positive cells differ in their susceptibility to lysozyme (an enzyme that attacks bacterial cell walls). The walls of gram-positive organisms are completely destroyed by lysozyme—the resultant structure, a cell with very little or no cell wall, is called a **protoplast**. The walls of gram-negative bacteria are more resistant and are not completely destroyed— the resultant cell, with a partial cell wall, is a **spheroplast**. Neither protoplasts nor spheroplasts are protected from osmotic lysis.

Bacteria of the genus *Mycoplasma* do not have cell walls.

L forms are bacteria that have defective cell walls.

Archaebacteria have cell walls that lack peptidoglycan and consist of pseudomurein.

Structures **interior** to the cell wall include the plasma membrane, the cytoplasm, and cytoplasmic constituents such as DNA, ribosomes, and inclusions.

The **plasma membrane** is a dynamic, selectively permeable membrane enclosing the cytoplasm. It is located between the cell wall and the cytoplasm, and it regulates movement of substances, including water, into and out of the cell. The plasma membrane consists of a phospholipid bilayer containing both integral and peripheral proteins. This type of membrane is called a **fluid mosaic** and is found in both prokaryotic and eukaryotic cells.

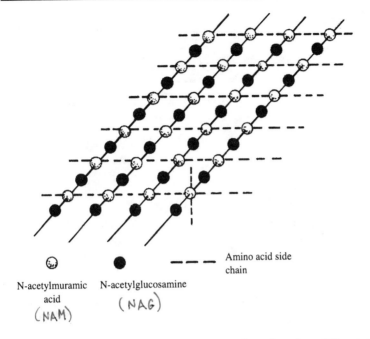

N-acetylmuramic acid
(NAM)

N-acetylglucosamine
(NAG)

Amino acid side chain

The molecules of peptidoglycan, above, show alternation of two different simple amino acid sugars that are held by amino acid side chains.

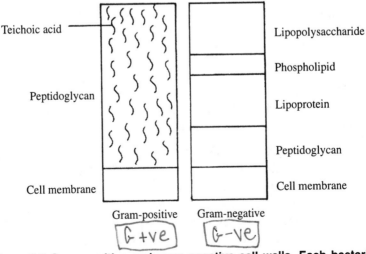

Teichoic acid

Peptidoglycan

Cell membrane

Lipopolysaccharide

Phospholipid

Lipoprotein

Peptidoglycan

Cell membrane

Gram-positive   Gram-negative

(G+ve)   (G-ve)

Figure 3.3 Gram-positive and gram-negative cell walls. Each bacterial cell wall differs by structure and type of peptidoglycan present.

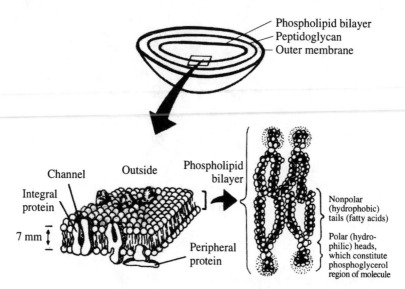

Figure 3.4 Fluid-Mosaic Model of the Plasma Membrane

The **DNA** of prokaryotes consists of a single, circular chromosome located in an area of the cytoplasm known as the nuclear area or nucleoid. Sometimes bacteria contain additional small DNA circles known as **plasmids**. Plasmids are not chromosomes and do not contain essential genetic information. They sometimes carry genes for antibiotic resistance and are a means for the spread of antibiotic resistance among bacterial populations.

Inclusions are accumulations of reserve deposits. Membrane-enclosed inclusions (**vesicles**) comprise carboxysomes, gas vacuoles, and lipid inclusions. Those without membranes are called **granules**.

Metachromatic granules are large inclusions, so called because they stain red with methylene blue. These granules are also referred to as **volutin granules** and are an indication of inorganic phosphate storage. These types of granules are commonly found in algae, fungi, protozoa, and bacteria. They are characteristic of *Corynebacterium diphtheriae*, the cause of diphtheria. **Polysaccharide granules** are reddish-brown or blue when stained with iodine. The reddish-brown stain is an indication of glycogen inclusions. The blue stain is an indication of starch inclusions. *Thiobacillus* (spp.) or "sulfur bacteria" obtain energy by oxidizing sulfur and contain **sulfur granules**.

During periods of adverse conditions, some bacteria form heat-resistant resting cells known as **endospores** (within the cell that formed them) or **exospores** (outside the cell that formed them). These dormant structures are highly resistant to heat, desiccation, chemical disinfection, stains, and radiation. Spores are not a means of reproduction in bacteria as they are in fungi; rather, a bacterial spore helps *the cell that produced it* to survive. Germination is the return to the vegetative state once conditions have improved. **Cysts** are resistant to drying, but are not resistant to heat.

Problem Solving Examples:

Q A biology student is presented with two jars. One is labelled "bacterial flagella" and the other is tagged "protozoan flagella." He is told they may be improperly labelled. How can he conclusively differentiate the flagella? Also discuss the role of flagella in bacteria.

A The extremely thin hairlike appendages that protrude through the cell wall of some bacterial cells are called flagella. The length of a flagellum is usually several times the length of the cell, but the diameter of a flagellum is only a fraction of the diameter (the width) of the cell. Bacilli and spirilla usually have flagella, which are rarely found on cocci. Flagella occur singly, in tufts, or equally distributed around the periphery of the cell.

Flagella are responsible for the motility of bacteria. Not all bacteria have flagella, and thus not all are motile. The mechanism by which the flagella move the bacterial cell is not completely understood; however, it is proposed that movement requires rotation of the semi-stiff flagellum. Whatever the mechanism, the flagellum moves the cell at a very high speed, and it enables the bacterium to travel many times its length per second.

Bacterial flagella are structurally different from the flagella of eukaryotic cells. This difference can be used to distinguish the bacterial (prokaryotic) flagella from the protozoan (eukaryotic) flagella. By examining a cross-section of the flagellum under an electron microscope, one can conclusively determine its identity. The eukaryotic flagellum contains a cytoplasmic matrix, with 10 groups of tubular fibrils embedded in the matrix: nine pairs of fibrils around the periphery of the eukaryotic flagellum surround two central single fibrils. Bacterial flagella consist of only a single fibril and lack the "9 + 2" structural organization of the eukaryotic flagellum. Most bacterial flagella have the same diameter as a single fibril from a eukaryotic flagellum and hence, they are usually thinner.

Q A biology student observes filamentous appendages on a bacterium and concludes they are flagella. However, he is told that the bacterium is nonmotile. He therefore concludes that the flagella are nonoperational. Why are both of his conclusions incorrect?

A It is rare for an organism to have a structure that serves no function. The filamentous appendages observed by the student are not flagella, but pili (or fimbriae). Pili are hairlike structures found on some gram-negative bacteria. They are smaller, shorter, and more numerous than flagella. They are found on nonmotile as well as motile bacteria, but do not function in motility.

There are several different kinds of pili. The sex pilus (singular) is involved in bacterial conjugation (the transfer of part of a chromosome from one bacterial cell to another). Most pili aid the bacterium in adhering to the surfaces of animal and plant cells, as well as to inert surfaces such as glass. The pili enable the bacteria to fix themselves to tissues from which they can derive nutrients.

Q Name an organism that produces spores in its life cycle and explain how this aids in its proliferation.

A While spores are primarily used as a means of reproduction in plants, there is one class of protozoans (the Sporozoa) that produces spores. These spores are highly resistant forms of the organism that can withstand extreme environmental conditions and are specialized for asexual reproduction.

The sporozoan *Plasmodium* is a spore-forming protozoan that causes the disease malaria. *Plasmodium* spores are produced initially by sexual reproduction in the body of a mosquito that has sucked the blood of an individual infected with the organism. These spores (sporozoites) may now be injected into another individual's bloodstream when the mosquito bites the new host. After invading the red blood cells, the spores reproduce asexually, producing many spores (merozoites) within each invaded cell. This causes the cell to burst, releasing the spores. Each is able to invade another red blood cell. It is thus through repeated sporulation that the organism proliferates in the bloodstream of the infected individual. Some merozoites develop into male and female gametocytes. It is these gametocytes that a mosquito sucks up when she bites. In the mosquito, the gametocytes develop into gametes. By sexual reproduction, the gametes fuse into zygotes, which become sporozoites, and the cycle repeats.

3.2.2 Bacterial Taxonomy

Important genera of the **spirochetes** include *Treponema, Borrelia,* and *Leptospira.* These are examples of gram-negative, aquatic animal parasites. They are distinguished by their long, slender shape and crawling movements due to an axial filament.

Members of the genera *Spirillum, Azospirillum, Campylobacter,* and *Bdellovibrio* are found in soil and aquatic environments. This group of **aerobic/microaerophilic, motile, helical/vibroid, gram-negative bacteria** include nitrogen-fixing bacteria.

Nonmotile or rarely motile gram-negative bacteria are not pathogenic. **Gram-negative, aerobic rods and cocci,** such as the genera *Pseudomonas, Legionella, Neisseria, Brucella, Bordatella, Francisella, Rhizobium, Agrobacterium, Acetobacter, Gluconobacter,* and *Zooglea,* are important organisms to medicine, industry, and the environment.

Many important pathogens, including Enterobacteriaceae *(Escherichia, Salmonella, Shigella, Klebsiella, Yersinia, Enterobacter),* Vibrionaceae *(Vibrio),* Pasteurellas *(Hemophilus, Gardnerella,* and *Pasteurella),* belong to the group of **facultative anaerobic, gram-negative rods.**

Escherichia coli (E. coli) is not usually considered a pathogen. However, it can be responsible for urinary tract infection, diarrhea, and very serious foodborne diseases.

The obligate anaerobes *Bacteroides* (spp.) and *Fusobacterium* (spp.) are part of the group of **anaerobic, gram-negative, straight, curved, or helical rods.** *Bacteroides* (spp.) are commonly found in the intestinal tract of humans and the rumens of ruminants, and *Fusobacterium* (spp.) are found in the gums.

Desulfovibrio is an important genus of the **dissimilatory sulfate-reducing or sulfur-reducing bacterial group.** This group of bacteria is important ecologically.

Veillonella (spp.) are nonmotile **anaerobic, gram-negative cocci.** These are associated with dental plaque.

Members of the genera **Coxiella,** as well as **Rickettsia** and **Chlamydia,** are important pathogens of arthropods and animals. *Coxiella burnetii* causes Q fever. Species of *Rickettsia* cause typhus and Rocky Mountain spotted fever. Certain species of *Chlamydia* cause trachoma, nongonococcal urethritis, psittacosis, and mild pneumonia.

Mycoplasmas lack cell walls; thus, they are penicillin resistant. They are parasites of animals, plants, and insects and are gram-negative bacteria. *M. pneumoniae* causes walking pneumonia. *Spiroplasma* (spp.) are parasites of plant-feeding insects. Species of *Ureaplasma* may be involved in urinary tract infections. *Thermoplasma* (spp.) are not pathogenic.

Gram-positive cocci include *Staphylococcus, Streptococcus, Lactococcus,* and *Enterococcus* and may be pathogenic. The most important species of *Staphylococcus* is *S. aureus,* which is the cause of toxic shock syndrome. *S. aureus* produces exotoxin, which increases this organism's pathogenicity. Streptococci cause diseases such as scarlet fever, pharyngitis, and pneumococcal pneumonia. Lactococci are important in the dairy industry, and enterococci are occasional pathogens.

Endospore-forming, gram-positive rods and cocci are important in industry and medicine. *Bacillus anthracis* is a rod-shaped member of this group and causes anthrax. Clostridia, also rod shaped, cause tetanus, botulism, diarrhea, and gas gangrene. *Sporosarcina* (spp.) are saprophytic soil cocci.

G+ve

Lactobacillus is a part of the group of **regular, nonsporing, gram-positive bacteria.** Lactobacilli have important uses in the food industry and are found in the human vagina, intestinal tract, and mouth. *Listeria monocytogenes* is a common contaminant of food products and is a pathogenic member of this group of bacteria.

G+ve

The bacterial group of **irregular, nonsporing, gram-positive rods** contains human pathogens, including *Corynebacterium, Propionibacterium,* and *Actinomyces. Corynebacterium diphtheriae* causes diphtheria. *Propionibacterium acnes* may be involved in acne. *Actinomyces israelii* causes actinomycosis.

G+ve

Mycobacteria include the important pathogens *Mycobacterium tuberculosis* and *M. leprae.* Members of this group are aerobic, nonsporing, nonmotile rods.

Some **nocardioforms** are aerobic, gram-positive bacteria. *Nocardia* and *Rhodococcus* are members of this group.

Members of the **gliding, sheathed and budding, and/or appendaged bacteria** include *Hyphomicrobium* and *Caulobacter.* Appendaged bacteria are characterized by the presence of a **prostheca** or protrusion. *Caulobacter* is an appendaged bacterium. Gliding, nonfruiting bacteria include *Cytophaga* and *Beggiatoa* and are characterized by their gliding motion. Gliding, fruiting bacteria include *myxobacterium. Hyphomicrobium* is a budding bacterium that does not reproduce by fission. *Sphaerotilus natans* are sheathed bacteria important in sewage treatment.

Nitrifying and sulfur-oxidizing aerobic **chemoautotrophic bacteria** include *Nitrosomonas, Nitrobacter,* and *Thiobacillus.* Members of this group are consequential in agriculture and in the environment.

G+ve,
G-ve

Archaebacteria, including *Methanobacterium, Halobacterium* and *Sulfolobus,* can be both gram-positive or gram-negative. Members of this group are characterized by the unusual environments in which they live. *Halobacteria* require high sodium chloride concentrations in their environment. *Sulfolobus* live at temperatures around 70°C, pH of 2, and in high sulfur concentrations. *Methanobacteria* are used in sewage treatment to produce methane from hydrogen and carbon dioxide.

Phototrophic bacteria are photosynthetic, gram-negative bacteria that use light as a source of energy. Green and purple sulfur bacteria are **anoxygenic** bacteria that do not produce oxygen during photosynthesis. *Anabaena* are cyanobacteria and are oxygenic phototrophic bacteria.

G-ve

Actinomycetes are filamentous, gram-positive bacteria, which include *Streptomyces, Frankia,* and *Micromonospora.* Some species of *Streptomyces* produce many antibiotics, which are used commercially.

G+ve

Problem Solving Examples:

Q How could rickettsias and chlamydiae be mistaken for viruses rather than bacteria?

A Bacteria, in general, are not intracellular parasites. However, rickettsias, chlamydiae, and viruses are obligatory intracellular parasites. In addition, rickettsias and chlamydiae are very small, and filtration cannot be used to remove these microorganisms from culture. Morphologically, the rickettsias and chlamydiae resemble bacteria more than viruses in that the rickettsias and chlamydiae have a plasma membrane and ribosomes. They reproduce by binary fission, possess both RNA and DNA, and are sensitive to antibiotics.

Q Why is *Gardnerella* classified more than once in *Bergey's Manual*?

A *Gardnerella* is classified as a facultatively anaerobic, gram-negative rod and as an irregular, nonsporing, gram-positive rod because these microorganisms can stain negative and positive by the Gram stain. The reason for the irregular staining patterns is due to the cell wall, which is structurally similar to gram-positive cells except that it is much thinner. The thin wall of *Gardnerella* causes the gram-negative staining pattern.

3.3   The Eukaryotic Cell

The most characteristic feature of a eukaryotic cell is the presence of membrane-bound organelles, especially the nucleus to which the name refers (*eu*—meaning true, and *karyon*—for nucleus).

Structures **external** to the cell wall or glycocalyx include flagella and cilia. Not all eukaryotic cells contain flagella or cilia. When present, eukaryotic flagella and cilia are composed of **microtubules**, which appear in a "nine pairs + two pairs" (9 + 2) arrangement. Eukaryotic flagella are composed of a protein called tubulin. Flagella are long and thin and few in number; cilia are short and numerous.

When present (in fungi and some plant cells), **eukaryotic cell walls** generally consist of either chitin or cellulose. The cell walls of yeast are composed of glucan and mannan.

Animal cells do not have a cell wall. The **glycocalyx** serves to strengthen the cell and provide a means of attachment.

In protozoans, other protective structures such as **pellicles** and **tests** may be produced.

The **eukaryotic plasma membrane** is similar to that of prokaryotes; i.e., it is composed of a phospholipid bilayer with associated pro-

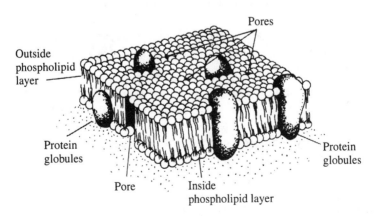

Outside phospholipid layer

Pores

Protein globules

Protein globules

Pore

Inside phospholipid layer

Figure 3.5 Schematic Depiction of a Typical Cell Membrane

teins (the **fluid mosaic** model). Additionally, the eukaryotic plasma membrane contains sterols and attached carbohydrates. (Sterols are absent in prokaryotes, with the exception of *Mycoplasma*.) The eukaryotic plasma membrane functions in the transport of molecules into and out of the cell.

The **cytoplasm** of a eukaryotic cell contains DNA, ribosomes, inclusions, organelles (e.g., nucleus, mitochondria, chloroplasts, Golgi apparatus), and a cytoskeleton.

The **DNA** of eukaryotes is arranged in paired chromosomes, with associated proteins (histones). It is located in a membrane-bound organelle called the nucleus.

There are two theories regarding the **evolution of cellular organelles**. According to the **endosymbiont theory**, organelles evolved from a symbiosis in which some prokaryotes lived inside other prokaryotes. Supporters of this hypothesis cite the occurrence of 70S ribosomes in both mitochondria and chloroplasts, and the fact that these organelles multiply by fission—both characteristics of prokaryotic cells. Eukaryotic cells contain larger ribosomes (80S) and undergo mitotic division.

According to the **autogenous hypothesis**, organelles evolved from the plasma membrane. Organelles that seem to support this theory include the nucleus and the Golgi apparatus.

Problem Solving Examples:

 According to the symbiont theory, what is the mechanism by which prokaryotes "took up residence" inside eukaryotic cells?

Many biologists believe that eukaryotic cells evolved when formerly free-living prokaryotes established a symbiotic relationship. Perhaps one prokaryotic cell ingested another, and instead of digesting it, the cell began to coexist with it. Much of the evidence for the symbiont theory involves similarities between bacteria and two of the main organelles of eukaryotic cells: chloroplasts and mitochondria.

What cell components typically found in a eukaryotic cell are missing from a bacterial (prokaryotic) cell?

Cells are classified as eukaryotic or prokaryotic. The former means "true nucleus," while the latter means "before the nucleus." Prokaryotes include all bacteria, including the cyanobacteria (blue-green algae), whereas eukaryotes include the protistan, fungal, animal, and plant cells. The most obvious difference between bacteria and eukaryotes is the absence of the nuclear membrane in bacteria. The DNA of the bacterial cell is generally confined to a specific area, but this region is not enclosed by a membrane. In addition, proteins, such as the histones of eukaryotic cells, are not found in association with bacterial DNA. Another distinctive feature is that during replication of the bacterial nuclear region, the mitotic spindle apparatus is not seen.

Bacterial cells also lack Golgi apparatus, mitochondria, and endoplasmic reticulum. This means that the ribosomes found in the cytoplasmic region are free, i.e., they are not bound to endoplasmic reticulum. Membranous systems are not completely absent in bacteria. The plasma membrane folds inward at various points to form mesosomes. These membranes may be involved with the origin of other intracellular structures and the compartmentalization and integration of biochemical systems. For example, although the electron transport system is located in the plasma membrane of bacteria, most of the respiratory enzymes are located in the mesosomes. The plasma membrane and mesosomes are the bacterial counterpart of the eukaryotic cells' mitochondria, serving to compartmentalize the respiratory enzymes. Those bacteria that contain chlorophyll do not contain plastids to "house" the chlorophyll. Instead, the chlorophyll is associated with membranous vesicles arising from mesosomes.

3.3.1 Algae

Algae is a catchall term for photosynthetic aquatic organisms, both unicellular and multicellular. Depending on its pigments and structures, an alga can be classified as either a plant or a protist. Algae are the primary energy producers in aquatic environments and are important in several food chains. They vary greatly in size (from microscopic to greater than 60 m in length), and also in shape and organization. Reproduction may be either sexual or asexual (fragmentation, division by splitting, unicellular spores).

3.3.1.1 Classification of Algae

Six algal divisions are recognized: the unicellular Chlorophyta, Chrysophyta, Euglenophyta, and Pyrrophyta, and the multicellular Phaeophyta and Rhodophyta.

Brown algae (Phaeophyta) or kelp can grow to be 50 meters in length. Brown algae also have very fast growth rates. Algin is harvested from the cell wall of brown algae and is used as a thickener in ice cream and cake decorations. It also has nonfood uses and is used in the production of rubber tires and hand lotion.

Red algae (Rhodophyta) have delicate branches called **thalli**. Agar is produced by red algae. Carrageenan is harvested from a red algae called Irish moss and is used as a thickener in evaporated milk and ice cream. Some may also be used medicinally as antihelminthics, as antidiarrheals, and in cancer chemotherapy.

Green algae (Chlorophyta) are thought to be the ancestors of terrestrial plants. Filamentous green algae form pond scum.

Diatoms (Chrysophyta) have complex cell walls composed of two halves that overlap like a petri dish containing pectin and silica and have distinctive patterns. In diatoms, energy is stored in the form of oil (crude oil). Deposits of dead diatoms form diatomaceous earth, which is used in detergents, abrasive polishes, and paint removers and as a filtering and insulating agent.

Dinoflagellates (Pyrrophyta) or **plankton** are free-floating organisms whose cell walls contain cellulose and silica. Eighty percent of the world's oxygen is produced by plankton. *Gonyaulax* is a dinoflagellate, which produces neurotoxins (saxitoxin), which cause paralytic shellfish poisoning. The shellfish are not harmed by the toxin, but it accumulates within the shellfish, making them highly toxic to vertebrates that eat them. These are also responsible for red tides, which are large blooms of dinoflagellates coloring the ocean red. Diatoms and other plankton are responsible for the supply of petroleum.

Euglenoids (Euglenophyta) are unicellular, flagellated organisms characterized by a **pellicle,** or semirigid plasma membrane; a **red eye spot (stigma),** or carotenoid organelle; a cytosome, used for digestion; and one or two **preemergent flagella** for locomotion.

Problem Solving Example:

Q All of the algae have the green pigment chlorophyll, yet the different phyla of algae show a great variety of colors. Explain.

A The pigments found in the different algae phyla are extremely varied, and their concentrations result in different colors. The earliest classifications of algae were based on color. Fortunately, later study of algae showed that algae of similar pigmentation also shared other important characteristics and that the older classifications were still valid.

In addition to the green pigment chlorophyll, most algae possess pigments of other colors called accessory pigments. The accessory pigments may play a role in absorbing light of various wavelengths. The energy of these light wavelengths is then passed on to chlorophyll. This absorption widens the range of wavelengths of light that can be used for photosynthesis.

Although there are several forms of chlorophyll found in eukaryotic plants, the most important forms for photosynthesis are chlorophyll a (absorption peak at 665 nm) and chlorophyll b (absorption peak at 645 nm). Chlorophyll c is found in some eukaryotic algae.

The accessory pigments phycocyanin and phycoerythrin serve this function in the red algae, the Rhodophyta. These pigments give the Rhodophyta their characteristic red color, although occasionally, they may be black. The red algae often live at great depths in the ocean. The wavelengths absorbed by chlorophyll do not penetrate to the depths at which the red algae grow. The wavelengths that do penetrate deep enough are mostly those of the central portion of the color spectrum. These wavelengths are readily absorbed by phycoerythrin and phycocyanin. The energy trapped by these pigments is then passed on to chlorophyll a, which utilizes this energy for photosynthesis.

In the green algae, the Chlorophyta, the chlorophyll pigments predominate over the yellow and orange carotene and xanthophyll pigments. The predominance of carotene pigments imparts a yellow color to the golden algae, members of the phylum Chrysophyta. The diatoms, the other important class of Chrysophyta, possess the brown pigment fucoxanthin. The Pyrrophyta (dinoflagellates) are yellow-green or brown, due to the presence of fucoxanthin and carotenes. Some red dinoflagellates are poisonous, containing a powerful nerve toxin. The blooming of these algae are responsible for the "red tides" that kill millions of fish.

The brown algae, the Phaeophyta, have a predominance of fucoxanthin. These algae range in color from golden brown to dark brown or black. The prokaryotic cyanophyta (blue-green algae) have the blue pigment phycocyanin as well as phycoerythrin, xanthophyll, and carotene. The Euglenophyta contain chlorophyll a and b and some carotenoids.

3.3.2 Fungi

The **fungi** (molds, yeast, and mushrooms) are nonmotile, nonphotosynthetic eukaryotes. They absorb nutrients from their environment. A few species are **parasitic**, but most are **saprophytic**, absorbing nutrients from dead organic matter. Fungi are important ecologically; they are, along with bacteria, the decomposers of the world. They are important economically as well, especially in the area of food spoilage. Fungal cell walls generally contain chitin, although yeast cell walls contain the complex polysaccharides glucan and mannan.

A **mushroom** consists of a vegetative structure known as a **thallus**, which is composed of filaments called **hyphae**. A **mycelium** is a mass of hyphae. **Molds** consist of long chains of cells that occasionally form hyphae; they reproduce by spores.

Yeasts are unicellular fungi that reproduce asexually by budding. Buds that do not separate from the mother cell form **pseudohyphae.** **Dimorphic** fungi appear yeastlike at 37°C, moldlike at 25°C. Some yeast have pili on their cell walls. These structures, similar to the pili of bacteria, may be involved in the sexual reproduction of yeast.

3.3.2.1 Classification of Fungi

Although fungi reproduce both sexually and asexually, they are divided into two groups, the **Ascomycetes** and the **Basidiomycetes,** on the basis of their type of sexual reproduction. In a third group of fungi, however, the **Fungi Imperfecti,** the sexual stage has never been observed.

Deuteromycota (imperfect fungi) are a phylum of fungi that have not yet been observed to produce sexual spores. Members of this phylum, therefore, cannot be classified. However, once sexual spores are observed, the organisms may be classified.

Zygomycota (conjugation fungi) include *Rhizopus nigricans,* the common black bread mold, which produces asexual sporangiospores and sexual **zygospores.** Zygospores are large, surrounded by a thick wall, and can remain dormant if the environment is too harsh for growth.

Ascomycota (sac fungi) produce **conidiospores,** which produce a spore dust when disturbed. The sexual spore is classified as an ascospore because the spores are produced in an ascus or saclike structure. *Eupenicillium* is a member of this phylum. *Claviceps purpurea* is a parasitic member of this class parasitizing rye and other grasses and causing the disease ergot.

Basidiomycota (club fungi) include mushrooms. These organisms produce **basidiospores** and asexual conidiospores. Basidiospores form on a base pedestal, the **basidium.** They are used as food and help to decompose plant debris.

3.3.2.2 Fungal Disease

Mycosis is the term for any **fungal infection** or **disease** and can include systemic mycoses, subcutaneous mycoses, cutaneous mycoses, and superficial mycoses. Fungi may be important in opportunistic infections of debilitated or immunosuppressed patients. Examples of opportunistic mycoses are mucormycosis, aspergillosis, candidiasis, and thrush. Mucormycosis is caused by *Rhizopus* and *Mucor.* Aspergillosis is caused by inhalation of *Aspergillus* spores. Candidiasis is caused by *Candida albicans.* Thrush causes an inflammation of the mouth and throat.

3.3.2.3 Commercially Important Fungi

Saccharomyces cerevisiae is a **yeast** important in the food industry. It is used in the production of bread and wine. *Trichoderma* produces the enzyme cellulase used in the production of fruit juice. *Taxomyces* is a **fungus** that produces taxol, an anticancer drug. *Phytophthora infestans* is the potato fungus responsible for the potato crop failure in Ireland in the 1800s. Dutch elm disease is caused by *Ceratocystis ulmi.*

Problem Solving Example:

How are fungi important to humans?

Fungi are divided into four groups: Oomycetes (Oomycota)—the egg fungi, Zygomycetes (Zygomycota)—zygosporeforming fungi, Ascomycetes (Ascomycota)—the sac fungi, and Basidiomycetes (Basidiomycota)—the club fungi.

Fungi are both beneficial and detrimental to humans. Beneficial fungi are of great importance commercially. Ascomycetes, or sac fungi, are used routinely in food production. Yeasts, members of this group, are utilized in liquor and bread manufacturing. All alcohol production relies on the ability of yeasts to degrade glucose to ethanol and carbon dioxide, when they are grown in the absence of oxygen. Yeasts used in alcohol production continue to grow until the ethanol concentration reaches about 13 percent. Wine, champagne, and beer are not concentrated any further. However, liquors such as whiskey or vodka are then distilled, so that the ethanol concentration reaches 40 to 50 percent. The different types of yeasts used in wine production are in part responsible for the distinctive flavors of different wines. Bread baking relies on CO_2 produced by the yeasts that cause the dough to rise. Yeasts used in baking and in the brewing of beer are cultivated yeasts and are carefully kept as pure strains to prevent contamination. Sac fungi of genus *Penicillium* are used in cheese production. They are responsible for the unique flavor of cheeses such as Roquefort and Camembert. The medically important antibiotic penicillin is also produced by members of this genus. Certain Ascomycetes are edible. These include the delicious morels and truffles.

The club fungi, or Basidiomycetes, are of agricultural importance. Mushrooms are members of this group. About 200 species of mushrooms are edible, while a small number are poisonous. The cultivated mushroom, *Agaricus campestris,* differs from its wild relatives and is grown commercially.

Fungi are often of agricultural significance in that they can seriously damage crops. Members of the Oomycetes, also known as water molds, cause plant seedling diseases, downy mildew of grapes, and potato blight (this was the cause of the Irish potato famine). Mildew is a water mold that grows parasitically on damp, shaded areas.

Rhizopus stolonifer, a member of the Zygomycetes, is known as black bread mold. Once very common, it is now controlled by refrigeration and by additives that inhibit mold growth. The Ascomycete *Claviceps purpurea* causes the disease ergot, which occurs in rye and other cereal plants and results in ergot poisoning of humans and livestock. This type of poisoning may be fatal. The disease caused in humans is called St. Vitus' dance. Visual hallucinations are a common symptom of this disease. Lysergic acid is a constituent of ergot and is an intermediate in the synthesis of LSD. The "dance macabre" of the Middle Ages is now believed to have been caused by ergot poisoning.

Basidiomycetes are also responsible for agricultural damage. Certain club fungi are known as smuts and rusts. Smuts damage crops such as corn, and rusts damage cereal crops such as wheat. Bracket fungi, another type of club fungi, cause enormous economic losses by damaging wood of both living trees and stored lumber.

Fungi are also important to humans because of the diseases they cause in humans and livestock. *Candida albicans* causes a throat and mouth disease, "thrush," and also infects the mucous membranes of the lungs and genital organs. Many skin diseases are caused by fungi, including ringworm and athlete's foot.

3.3.3 Lichens and Slime Molds

The lichens and slime molds are not easily classified.

A **lichen** is the result of a symbiosis between an alga and a fungus. The fungus provides structural support and the alga provides nutrients. Lichens are able to occupy habitats that would not be suitable for either the fungus or the alga alone.

Slime molds are sometimes classified as fungi, sometimes as protozoans. Most are saprophytic, but some are parasitic. They exist in two forms: cellular and acellular.

An **acellular,** or **plasmodial,** slime mold is a multinucleated mass of protoplasm that engulfs bacteria and organic matter as it moves along.

Cellular slime molds ingest bacteria by phagocytosis; they resemble amoebas.

Problem Solving Examples:

 In what ways are slime molds like true fungi? In what ways do they resemble animals?

Both cellular slime molds (Acrasiomycota) and true slime molds (Myxomycota) have unusual life cycles containing fungus like and animallike stages. Slime molds have membrane-bound nuclei, are heterotrophic, ingest food, lack cell walls, and produce fruiting bodies. They belong to the kingdom Protista in the phylum Gymnomycota.

The true slime mold's adult vegetative stage is decidedly animal-like. At this stage, the slime mold is a large, diploid, multinucleated amoeboid mass called a plasmodium. It moves about slowly and feeds on organic material by phagocytosis. Plasmodium growth continues as long as an adequate food supply and moisture are available. When these run short, the plasmodium becomes stationary and develops organs specialized to produce haploid spores, known as the fruiting bodies. At this stage, the true slime mold is similar to the fungi. Meiosis occurs in the fruiting body, and spores with cellulose cell walls are released. Spores are a resistant and dormant form of a slime mold. When the spores germinate, they lose their cell walls and become flagellated gametes. Gametes fuse to become zygotes. The zygotes lose their flagella, become amoeboid, and grow into multinucleated plasmodial slime molds.

Cellular slime molds differ from the true (acellular) slime molds in being haploid and in that the amoeboid cells, on swarming together, retain their identity as individual cells. The cellular slime molds resemble amoebas throughout most of their life cycle: they lack cell walls, move about, and ingest particulate matter. Under certain conditions, the amoebas aggregate to form a multicellular pseudoplasmodium, called a slug. The pseudoplasmodium becomes stationary, and fruiting bodies are formed. Spores are not produced by cellular division, but by the formation of cell walls around the individual amoeboid cells. Each spore becomes a new amoeboid cell when it germinates. Cellular slime molds are haploid throughout their life cycles. The formation of a fruiting, spore-forming body is characteristic of fungi.

 What are lichens? Describe the relationship that exists in a lichen. How do they reproduce?

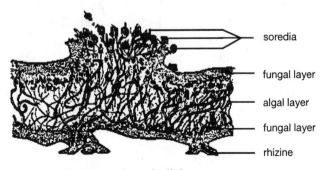

Cross section of a lichen

A Lichens are composite organisms consisting of algae and fungi. They grow on tree bark, rocks, and other substrates not suitable for the growth of plants. Lichens may be found in low-temperature environments characteristic of polar regions and very high altitudes.

Structurally, a lichen can be likened to a fungal "sandwich" whose hyphae entwine a layer of algal cells (see figure). Structures known as rhizoids, which are short twisted strands of fungal hyphae, serve to attach the bottom layer to the substrate.

Not all species of algae or all species of fungi can enter into a lichenlike relationship. Most of the fungi found in lichens are Ascomycetes, although there are a few of the Basidiomycetes. Algae of the lichens are from the Cyanophyta (blue-green algae) or the Chlorophyta (green algae) phylum. Each lichen thallus consists of a single species of fungus associated with a single species of alga.

There are two types of lichens: fruticose and crustaceous. The fruticose have an erect, shrublike morphology. The crustaceous are very closely attached to the substrate or may even grow within its surface.

Lichens are the product of a relationship called symbiosis, in which each partner of the association derives something useful from the other for its survival. The alga provides the fungus with food, especially carbohydrates produced by photosynthesis, and possibly vitamins as well. The fungus absorbs, stores, and supplies water and minerals required by the alga, as well as providing protection and a supporting framework for the alga.

Lichens appear to reproduce in a variety of ways. Fragmentation may occur, in which bits of the thallus are broken off from the parent plant and produce new lichens when they fall on a suitable substrate. Lichens may produce "reproductive bodies" called soredia, which are knots of hyphae containing a few algal cells. In addition, the algal and fungal components of a lichen may reproduce independently of each other. The fungal component produces ascospores, and the algal component reproduces by cell division or infrequently via sporulation. Some species of lichens in the Arctic have been alive for 4,500 years, which suggests a very well-balanced association between the symbionts. Lichens grow very slowly because of their low metabolic rate. They are also very resistant to heat and desiccation.

Lichens produce many interesting organic products. Unusual fats and phenolic compounds make up from two to twenty percent of the dry weight of the lichen body. Litmus, the pigment indicator, and essential oils used in perfumes are obtained from lichens.

3.3.4 Protozoans

Protozoans are unicellular eukaryotes. Most are motile, microscopic, and heterotrophic, feeding on bacteria or particulate organic matter. There are thousands of species, about two dozen of which cause disease in humans.

Lacking a cell wall, protozoans employ such protective structures as **pellicles** (a strong outer covering found in ciliates and some amoebas), **tests** (shells of calcium carbonate or silica), and **trichocysts** (specialized defense organelles). Some protozoans also form **cysts**, resting stages with thick resistant coverings.

Trophozoite—the active feeding stage of a protozoan (the term is used for both free-living and parasitic protozoans).

Protozoans reproduce **sexually** by conjugation, autogamy, or syngamy; and **asexually** by budding, binary fission, multiple fission, or plasmotomy.

Conjugation involves the transfer of DNA between cells via a temporary connection.

Autogamy is a modification of conjugation.

Syngamy is the fusion of two different types of sex cells.

Plasmotomy is splitting into two or more multinucleated cells.

Problem Solving Example:

Q While encystment is not, in most cases, a means of reproduction, still it is a means of continuation of the species. Explain how encystment is used, using a cyst-forming organism as illustration.

A Encystment is used by a number of protozoans that live in the water. Should their pond become desiccated or uninhabitable, these organisms would develop hard, protective coatings, and the organisms would go into a state of dormancy. In this encysted state they can withstand extremely unfavorable conditions. When favorable conditions return (such as clean water), the wall of the cyst softens and the organisms escape.

Some pathogenic protozoans develop cysts that play a role in the spread of infection. *Entamoeba histolytica* is a parasitic protozoan that lives in the lining of the human large intestine, causing the disease amoebic dysentery. This parasite secretes a tissue-dissolving enzyme that causes ulceration of the large intestine. Some of these protozoans die when they pass from the body into the feces, but some of these organisms develop cysts and continue to live in dry, unfavorable conditions outside the body for a long time. Should such cysts enter the body of another person, the wall of the cysts would dissolve, and the parasites would escape into the new host's body. A new infection would thus begin.

3.3.4.1 Classification of Protozoans

Protozoans are classified according to their mode of locomotion. Those using whiplike **flagella** for locomotion belong to the Mastigophora; those using hairlike **cilia** are the Ciliates; those moving via **pseudopodia** (amoeboid movement) belong to the Sarcodina; and those lacking a means of locomotion as adults belong to the Sporozoa (all sporozoans are obligate parasites).

Sarcomastigophora is the phylum of protozoans that consists of organisms that use either pseudopods or flagella as a form of locomotion. For example, *Naegleria* and other species use both forms of locomotion.

Sarcodina is the subphylum of the amoeboflagellates that use pseudopods or cytoplasmic projections to move. *Entamoeba histolytica* causes amoebic dysentery, which is endemic in many parts of the world. Some amoebas, such as *Acanthamoeba,* can pass through the blood-brain barrier.

Some parasitic amoebae that use flagellae are called **flagellates.** Among them are *Giardia lamblia,* which causes intestinal infections, *Trichomonas vaginalis,* which causes a genitourinary infection, and *Trypanosoma,* which causes African sleeping sickness *(T. brucei gambiense)* and Chagas' disease *(T. cruzi).* This phylum of flagellates is **Mastigophora.**

Ciliates use cilia for locomotion (Ciliophora) and to propel food towards the mouth. Ciliates are **multinucleate** with both a macronucleus and one or more micronuclei. A parasitic ciliate is *Balantidium coli,* which causes dysentery.

Apicomplexa is the phylum that consists of mature, nonmotile organisms with complex life cycles. *Plasmodium,* which causes malaria, is an example of an **apicomplexan.** *Toxoplasma gondii,* which spends its life cycle in domestic cats, can cause congenital infections in pregnant women.

Cryptosporidium causes respiratory and diarrheal diseases in immunosuppressed patients. *Pneumocystis carinii* causes pneumonia.

Microspora is a phylum of protozoans that are obligatory intracellular parasites responsible for chronic diarrhea and keratoconjunctivitis. The human pathogen is *Nosema* and is commonly found in AIDS patients. *Nosema locustae* is used in residual control of rangeland grasshoppers.

Problem Solving Example:

 What are the chief characteristics of the protozoans?

The protozoans are a heterogeneous assemblage of a large number of species, which are almost exclusively microscopic organisms.

The protozoans are grouped into the phylum Protozoa within the Animal Kingdom, although this classification remains controversial. Some biologists still believe that the protozoans have more in common with the Kingdom Protista. Protozoans live either singly or in colonies. These organisms are usually said to be unicellular. Therefore, they contain no tissue or organs, which are defined as aggregations of differentiated cells. Instead of organs, they have functionally equivalent subcellular structures called organelles. These organelles do show a great deal of functional differentiation for the purposes of locomotion, food procurement, sensory reception, response, protection, and water regulation. Certain protozoans have interesting plantlike characteristics in both structure and physiology.

Reproduction among the protozoans is variable. An individual may divide into two, usually equal halves, after which each grows to the original size and form. This form of reproduction is called binary fission and can be seen in the flagellates, among the ciliates, and in organisms such as the amoeba. Multiple fission, or sporulation, where the nucleus divides repeatedly and the cytoplasm becomes differentiated simultaneously around each nucleus resulting in the production of a number of offspring, is also seen among the protozoans. Another type of reproduction characteristic of the protozoans is plasmotomy, which is the cytoplasmic division of a multinucleate protozoan without nuclear division, resulting in smaller multinucleate products.

Budding is another reproductive process by which a new individual arises as an outgrowth from the parent organism differentiating before or after it becomes free. All the reproductive mechanisms thus far mentioned illustrate asexual means of reproduction. Sexual reproduction may also occur by the fusion of two cells, called gametes, to form a new individual, or by the temporary contact and nuclear exchange (conjugation) of two protozoans (for example, two *Paramecia*). The result of conjugation may be "hybrid vigor," defined as the superior qualities of a hybrid organism over either of its parental lines. Some species have both sexual and asexual stages in their life cycles.

With regard to their ecology, protozoans are found in a great variety of habitats, including the sea, fresh water, soil, and the bodies of other organisms. Some protozoans are free-living, meaning that they are free-moving or free-floating, whereas others have sessile organisms. Some live in or upon other organisms in either a commensalistic, mutualistic, or parasitic relationship.

The mechanism for the acquisition of nutrition is also variable among the protozoans. Some are holozoic, meaning that solid foods such as bacteria, yeasts, algae, protozoans, and small metazoans or multicellular organisms are ingested. Others may be saprozoic, wherein dissolved nutrients are absorbed directly; holophytic, wherein manufacture of food takes place by photosynthesis; or mixotrophic, which uses both the saprozoic and holophytic methods.

It should be pointed out that the unicellular level of organization is the only characteristic by which the phylum Protozoa can be described. In all other respects, such as symmetry and specialization of organelles, the phylum displays extreme diversity.

3.3.5 Animal Microbes—Helminths

Certain stages in the life cycles of **helminths** are microscopic. Helminths include the Platyhelminthes, or flatworms (e.g., the liver fluke), and the nematodes, or roundworms (e.g., hookworm).

3.3.5.1 Classification of Helminths

The phylum **Platyhelminthes (flatworms)** contains the trematodes and cestodes. Members of this phylum have incomplete digestive systems. These flatworms have only one opening through which both nutrients and waste products flow.

The **trematodes** are **flukes** characterized by a ventral or oral sucker by which the organism holds on to its host and by a **cuticle,** or non-living covering, through which it absorbs food. *Clonorchis sinensis* is an Asian liver fluke and causes clonorchiasis. *Schistosoma* is a fluke that infects humans through the skin and causes the disease schistoso-miasis.

Paragonimus westermani is a lung fluke. The adult fluke lives in the lung and lays eggs in the sputum of the bronchi. Sputum is swallowed and eggs are excreted by the human host. The eggs must then enter water to complete the life cycle. The **miracidium** develops within the egg. The secondary hosts are snails that live in the water. In the snail the miracidium develops into a **redia,** which produces **rediae** by asexual reproduction. The rediae develop into **cercaria,** which then find a tertiary host, a crayfish. **Metacercaria** develop from cercaria in the crayfish. When the crayfish are eaten by humans, the metacercaria find their way to the lungs, where they develop into an adult fluke.

Cestodes are **tapeworms.** These are intestinal parasites, which may attach to their host by sucker or by hook. The digestive system is absent in these organisms. The head is called the **scolex,** which contains the sucker or hook, and the body consists of segments called **proglottids.** Each body segment is hermaphroditic.

Taenia saginata is the beef tapeworm, whose adult form lives in the human intestine and can reach lengths of 6 meters. Their eggs are ingested by grazing cattle. The eggs hatch in the bovine intestine, and the larval form migrates to the muscle where these encyst as **cysticerci.** The cysticerci can be ingested by humans when they consume the contaminated meat.

Taenia solium is the porcine tapeworm. The hosts of this tapeworm are also humans. Cysticercosis is a disease characterized by human-to-human transmission of eggs, where the cysticerci encyst in the brain of the human host.

The phylum **Nematoda,** or **roundworms,** consists of organisms with complete digestive systems. Some nematodes are free living in soil while others are parasites and require a host. Male roundworms have one or two spicules on their posterior ends that are used to guide the male sperm to the genital pore on the female. There are two types of roundworms, those whose eggs are infective and those whose larvae are infective.

Enterobius vermicularis, or pinworms, may complete their entire life cycle within a human host. Eggs are laid on the perianal skin of the host and are transmitted through clothing or bedding. The adult form of *Ascaris lumbricoides* inhabits the intestines of humans, pigs, and horses. The eggs are transmitted through feces and are ingested by hosts.

Necator americanus is the adult hookworm. The adult form is found in the small intestine of humans, and eggs are found in feces. The eggs hatch in soil, and the larval form may infect a host via the skin. *Trichinella spiralis,* which causes trichinosis, is transmitted via undercooked pork meat. The larval form is encysted in this meat. The female does not lay eggs; instead, eggs develop within the female. The female gives live birth.

Problem Solving Example:

Q The development of a parasitic mode of existence is a remarkable example of an adaptation that has evolved to permit one organism to exist at the expense of another. Among the flatworms, the flukes are parasitic. Describe the body structure of a fluke, its modifications for a parasitic existence, and its life cycle.

A The adult fluke is roughly one inch long, and its body is quite flat. The worm is covered by a cuticle that is secreted by the underlying cells. The cuticle is one of its adaptations for parasitism, for it protects the worm from the enzymatic action of the host. With the development of the cuticle, the cilia and sense organs so characteristic of other flatworms have disappeared. The mouth is located at the anterior end and is surrounded by an oral sucker. Another sucker, the ventral sucker, is located some distance posterior to the mouth. The ventral sucker is used to attach the fluke to the body of its victim, the host. The anterior sucker, with the aid of the muscular pharynx, is used to withdraw nutrients from the host. The mouth opens into a muscular pharynx, and just in back of this, the alimentary tract branches into two long intestines that extend posteriorly almost the entire length of the body.

By far the most complicated aspect of these worms is their reproductive cycle. The fluke is hermaphroditic; that is, it contains both male and female sexual organs. There is no copulatory structure in the flukes, and reproduction occurs through self-fertilization. Following fertilization, the eggs are expelled from the fluke, which at this point is within the host's body; they then exit from the host's body in the feces. Hatching of the eggs in the oriental liver fluke occurs only when the eggs are eaten by certain species of freshwater snails, and takes place in the digestive system of the snail. Here, the egg hatches into a ciliated, free-swimming miracidium. Within the digestive system of the snail, which is called the first intermediate host, the miracidium develops into a second larval stage, called a sporocyst. Inside the hollow sporocyst, germinal cells give rise to a number of embryonic masses. Each mass develops into another larval stage called a redia, or daughter sporocyst. Germinal cells within the redia again develop into a number of larvae called cercariae. The cercaria possesses a digestive tract, suckers, and a tail. The cercaria is free-swimming and leaves the snail. If it comes in contact with a second intermediate host, an invertebrate or a vertebrate, it penetrates the host and encysts. The encysted stage is called a metacercaria. If the host of the metacercaria is eaten by the final vertebrate host, the metacercaria is released, migrates, and develops into the adult form within a characteristic location in the host, usually the bile passages of the liver. A human, the final host, usually gets the fluke by eating a fish that contains the encysted cercaria.

3.4   Viruses

Depending on one's point of view, viruses are either extremely complex nonliving entities or extremely simple forms of life. Capable of parasitizing every kind of life, they have no nutritional patterns of their own. They do not grow. They have no cellular structures. Their only observable activity is nucleic acid replication, and this can be accomplished only within a living cell. They are not assigned to any kingdom.

The mature virus particle—the **virion**—consists of nucleic acid (either RNA or DNA) and a protein sheath (the **capsid**), and can transfer its nucleic acid among host cells. It may or may not have an **envelope**. Enveloped viruses may have spikes that project from the surface and appear to be involved in the process of viral attachment to host cells.

The viral life cycle consists of both intra- and extracellular phases. The usual sequence of events is **attachment** to the host cell, **penetration** of the host cell, **uncoating** of the virion, **manufacture** of viral parts (under direction of the viral genome but carried out using the host's machinery, proteins, and energy), **assembly** of new virus particles, and **release** of new viruses.

Virus classification is based on morphology (including size and shape), structure, chemical and physical characteristics (especially nucleic acid chemistry and capsid organization), mode of replication, and host range.

Viruses are very small: 20–300 nm.

The major viral shapes can be described as **icosahedral** (e.g., herpes and polio viruses), **helical** (e.g., rabies, tobacco mosaic), and **complex** (see Figure 3.6).

Host range—the range of host cells in which a virus can multiply. Categories include animal viruses (e.g., chicken pox and smallpox in humans, rabies in dogs), plant viruses (e.g., tobacco mosaic virus), bacterial viruses (bacteriophages), or cyanophyte viruses (cyanophages). A particular virus can usually infect only a few species within the more general host categories.

Animal viruses may contain either RNA or DNA; however, most plant viruses contain RNA. **Plant viruses**, usually polyhedral or helical, can produce both internal and external effects and have an economic impact on agriculture.

Viroids are extremely small, low molecular weight RNA, infectious agents found mainly in plant diseases. Other so-called "small RNAs" involved in disease include satellite viruses, satellite RNAs, and defective interfering particles.

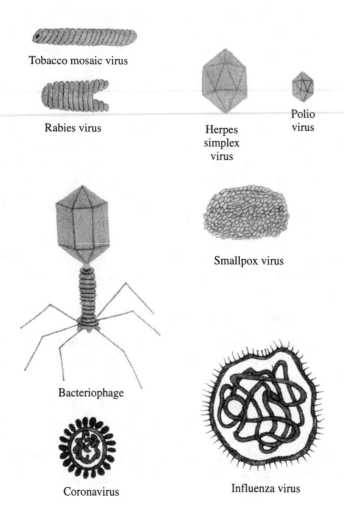

Tobacco mosaic virus

Rabies virus

Herpes
simplex
virus

Polio
virus

Smallpox virus

Bacteriophage

Coronavirus

Influenza virus

Figure 3.6 Shapes of Viruses

Problem Solving Examples:

 In what respects are viruses living things? How are they unlike
living things?

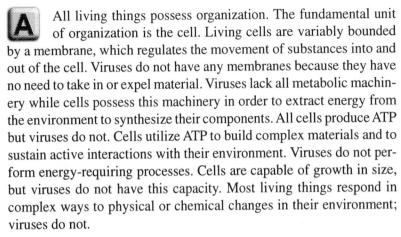

All living things possess organization. The fundamental unit of organization is the cell. Living cells are variably bounded by a membrane, which regulates the movement of substances into and out of the cell. Viruses do not have any membranes because they have no need to take in or expel material. Viruses lack all metabolic machinery while cells possess this machinery in order to extract energy from the environment to synthesize their components. All cells produce ATP but viruses do not. Cells utilize ATP to build complex materials and to sustain active interactions with their environment. Viruses do not perform energy-requiring processes. Cells are capable of growth in size, but viruses do not have this capacity. Most living things respond in complex ways to physical or chemical changes in their environment; viruses do not.

Most importantly, living things possess the cellular machinery necessary for reproduction. They have a complete system for transcribing and translating the messages coded in their DNA. Viruses do not have this system. Although they do possess either RNA or DNA (cells possess both), they cannot independently reproduce, but must rely on host cells for reproductive machinery and components. Viruses use their host's ribosomes, enzymes, nucleotides, and amino acids to produce the nucleic acids and proteins needed to make new viral particles. They cannot reproduce on their own. However, viruses are unlike nonliving things in that they possess the potentiality for reproduction. They need special conditions, such as the presence of a host, but they are able to duplicate themselves. Since nonliving things do not contain any nucleic acids, they cannot duplicate themselves.

This is the critical argument of those who propose viruses to be living things.

The difficulty in deciding whether viruses should be considered living or nonliving reflects the basic difficulty in defining life itself.

We know that radiation and chemicals can cause cancer. Can viruses be a cause of cancer?

Cancer involves an abnormality in the control of cell division and cell function. The factors that give cancer cells their characteristic property of unregulated growth are passed on from parent to progeny cancer cells. For this reason, it has been suggested that genetic changes occur within chromosomal DNA. These changes lead to a heritable cancer phenotype. Somatic mutations (mutations in cells that are not destined to become gametes) may cause a cancer, if the mutation upsets a normal device controlling cell regulation.

In contrast to the somatic mutation proposal is the hypothesis that most cancers are induced by viruses. Viruses that produce tumors in animals are called oncogenic viruses. While somatic mutations are suspected to cause a loss of functional genetic material, viruses are vehicles into a cell: they introduce new genetic material that may transform the cell into a cancerous type. In addition to multiplying and lysing their host cell, certain viruses can insert their chromosomes into the host chromosome. In some animal species, this process can transform the cell into a morphologically distinguishable cancer cell. At this point, the virus is in the prophage state. Therefore, absence of detectable viruses does not provide ample evidence for or against a viral cause of a cancer.

In 1911, an RNA virus, called the Rous sarcoma virus, was shown to be the causal agent of a sarcoma (a tumor of connective tissue) in chickens. Other RNA viruses cause sarcomas and leukemias (uncontrolled proliferation of leukocytes) in both birds and mammals. One group of DNA viruses causes warts on the skin of mammals. Other DNA tumor viruses are a mouse virus called polyoma and a monkey virus called SV 40. A Herpesvirus, EB (Epstein-Barr) causes infectious mononucleosis and is probably in Burkitt's lymphoma, a cancer prevalent in humid tropical regions. Research is being conducted to find viral causes of cancer in humans.

Q If DNAase is added to a bacterial cell, the DNA is hydrolyzed, the cell cannot make any more proteins and eventually dies. If DNAase is added to RNA viruses, they continue to produce new proteins. Explain.

A The infective ability of viruses is due to their nucleic acid composition. An individual virus contains either DNA or RNA but not both as is true for cells. Therefore, since the virus proposed in the question contains RNA, it is not affected by DNAase. The RNA replicates, forming a complementary RNA strand that acts as messenger RNA in order to code for the synthesis of new viral proteins. To produce new viral RNA, the viral RNA first synthesizes a complementary strand and thus becomes double-stranded. The double-stranded RNA serves as a template for synthesis of new viral RNA. The virus could have been tobacco mosaic virus, influenza virus or poliomyelitis virus. These are all viruses that contain single-stranded RNA as their genetic material. There are at least two groups of RNA viruses in which the RNA is normally double-stranded and assumes a double-helical form. DNA viruses, such as the smallpox virus, SV 40 (a tumor-inducing virus), and certain bacterial viruses, such as bacteriophages T_2, T_4, and T_6, contain double-stranded DNA. Yet there are some bacteriophages that have a single-stranded DNA molecule. It does not matter whether the genetic information is contained in DNA or RNA, or if it exists as a single strand or as a double helix. For viruses, the important point is that the genetic message is present as a sequence of nucleotide bases.

Q Viruses do not have any ribosomes which are the structures needed for protein synthesis. How are they able to synthesize their protein coats during their replication? In your explanation, describe the life cycle of a typical virus.

A Since viruses have no independent metabolic activity, they are incapable of reproduction by fission, budding, or other simple means. Instead, they replicate by inserting their nucleic acid into a functional host cell. The host cell provides both the energy and structural machinery necessary for the production of new viral components. After these components are assembled, they are released from the host cell, and a new cycle of replication begins.

3.4.1 Bacteriophages

Bacteriophages contain either RNA or DNA. Some have a taillike structure through which they inject their nucleic acid into the bacterial host cell.

Bacteriophage infection follows one of two courses—lysis or lysogeny. If the infecting virus multiplies within the host cell and destroys it, the virus is said to be **lytic,** or **virulent.** On the other hand, if the virus does not replicate but rather integrates into the bacterial chromosome, the virus is said to be **temperate,** or **lysogenic.** The phage in the lysogenic cycle can spontaneously become lytic. The presence of the integrated virus, which is called a **prophage,** generally renders the cell resistant to infection by similar phages. Lysogeny does not result in the destruction of the host cell.

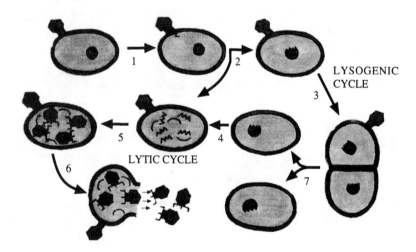

1. Phage attaches to host cell wall
2. Viral DNA is injected into cell
3. Viral DNA attaches to bacterial DNA
4. Viral DNA takes over cell and starts to reproduce
5. New protein coats are synthesized
6. Bacterium lyses and new viruses are released
7. Bacterial cell divides, producing more cells with viral DNA

Figure 3.7 Lysis and Lysogeny

Problem Solving Example:

Q A particular type of bacteriophage attaches to and penetrates a number of bacterial cells. After 20 minutes, some bacterial cells lyse and release many new viruses. Other bacteria remain intact and reproduce normally. After exposure to ultraviolet light, the remaining bacteria lyse within an hour. Explain.

A Viruses such as the T_2 bacteriophage always multiply when they enter a host bacterial cell. Lysis eventually occurs and the newly synthesized progeny are released from the cell. Some viruses such as the λ (lambda) phage do not always multiply and lyse their host. They are referred to as the temperate phages, in contrast to the virulent phages, which always kill their host. The temperate virus must be present in some inactive state within the bacterial cell. In this special relationship between the virus and the bacterium, called lysogeny, the viral DNA is incorporated into a specific section of the host's chromosome. The viral chromosome becomes an integral part of the host chromosome and is duplicated along with it during each cell division. The phage DNA can be transferred from one cell to another during bacterial conjugation. When the viral chromosome is integrated within the host chromosome, it is called the prophage. Bacteria containing prophages are called lysogenic bacteria, and viruses whose chromosomes can become prophages are called lysogenic viruses (in contrast to lytic viruses).

The prophage can remain integrated within the bacterial chromosome for many generations. Under certain natural conditions, the viral DNA becomes active and is removed from the host's chromosome. Once independent of the host chromosome, the lysogenic virus replicates and eventually lyses the cell. This entrance into the lytic cycle can be artificially induced by ultraviolet light or certain chemicals.

It is thought that while the viral DNA is inserted as a prophage, the transcription (the process where DNA is used to code for a complementary sequence of bases in an RNA chain, i.e., production of mRNA) of almost all phage genes is blocked. The only protein made by the λ phage DNA is the λ repressor, which causes the blockage of viral-specific mRNA transcription. Since no viral components can be made, the viral DNA remains inactive within the host chromosome. However, when the λ repressor is inactivated by an agent such as UV light, mRNA transcription occurs and the virus begins replication. Lysis eventually results with the release of new λ phages.

When a lysogenic virus is in the prophage state, its genes can produce phenotypic effects on the host bacterium. It can modify the cell wall, affect the production of enzymes and antigens, and possibly confer toxigenicity upon the lysogenic bacteria. Certain lysogenic phages enable the bacterium *Corynebacterium diphtheriae* to produce the toxin that causes the disease diphtheria.

Quiz: Survey of Microorganisms

1. Select the nonpathogenic bacterium.

    (A) *Clostridium*

    (B) *Escherichia*

    (C) *Salmonella*

    (D) *Staphylococcus*

    (E) *Treponema*

2. Which of the following is a single bacterial cell?

    (A) *Diplobacillus*

    (B) *Gonococcus*

    (C) *Staphylococcus*

    (D) *Streptococcus*

    (E) *Streptomyces*

3. Select the disease caused by a protozoa.

    (A) Chicken pox

    (B) Common cold

    (C) Malaria

    (D) Measles

    (E) Smallpox

4. *Neisseria gonorrhoeae* occurs as pairs of spherical organisms. Its morphological characteristic would be that of

    (A) cocci.

    (B) filamentous.

    (C) bacilli.

    (D) spiral.

    (E) None of the above.

5. All of the following are terms that describe viruses except

    (A) free-living.

    (B) host-dependent.

    (C) noncellular.

    (D) protein and nucleic acid makeup.

    (E) ultramicroscopic.

6. A bacteriophage

    (A) is a bacterium that phagocytoses other organisms.

    (B) is a bacterium that becomes phagocytosed by other organisms.

    (C) is a virus that infects bacteria.

    (D) is a fragment of DNA.

    (E) lives in a lysogenic cell.

7. A protozoan-induced disease is

    (A) acquired immune deficiency syndrome.

    (B) African sleeping sickness.

    (C) common cold.

    (D) polio.

    (E) spinal meningitis.

8. Viruses attack the bacterial cells as well as eukaryotic cells. The protein coat, or capsid, of a virus encloses a core of

    (A) RNA or DNA.

    (B) DNA only.

    (C) RNA only.

    (D) RNA and DNA.

    (E) None of the above.

9. Which of the following correctly describes the polyoma virus?

    (A) It is a DNA tumor virus.

    (B) When a permissive cell is infected, a transformation event may occur in one out of every 10^4 to 10^5 cases.

    (C) An RNA-DNA hybrid is formed from viral RNA and host deoxyribonucleotides.

    (D) It contains a reverse transcriptase.

    (E) None of the above.

10. Rickettsias

    (A) cause typhus and Rocky Mountain spotted fever in humans.

    (B) have no cell walls.

(C) exist as dense sporelike cells called elementary bodies.

(D) are endospore-forming, anaerobic, gram-positive rods.

(E) are peritrichously flagellated, gram-negative facultatively anaerobic rods.

ANSWER KEY

1.	(B)	6.	(C)
2.	(B)	7.	(B)
3.	(C)	8.	(A)
4.	(A)	9.	(A)
5.	(A)	10.	(A)

CHAPTER 4

Microbial Metabolism

4.1 General Terms

All of the chemical reactions that take place in a living organism, including *anabolism* and *catabolism,* are called the **metabolism**. As part of the metabolism, **anabolism/anabolic reactions** are reactions in which energy is used to synthesize complex organic molecules from simpler molecules. These reactions require an input of energy. **Catabolism/catabolic reactions** are reactions used by organisms to obtain energy, in which complex organic molecules are oxidized and broken down into simpler molecules. These reactions release energy.

Amphibolic pathways are sequences of metabolic reactions that integrate both anabolism and catabolism. Such a pathway may either capture energy or produce substances needed by the cell.

Some reactions require a catalyst. A **catalyst** is a substance that lowers the energy of activation for a reaction. The **energy of activation** is the energy required for a reaction to take place.

4.2 Enzymes

Enzymes are proteins that catalyze chemical reactions. They are specific—generally catalyzing only a single reaction or a group of related reactions. They are named according to their substrate (the compound they act upon) and function, e.g., glucose-6-phosphate dehydrogenase is an enzyme that functions as a dehydrogenase with glucose-6-phosphate as its substrate. Names of enzymes always end with *-ase*. Like all other proteins, enzymes can be destroyed by high temperatures; such **denatured** enzymes can no longer function as catalysts.

Enzyme activity (and thus, reaction rate) is influenced by pH, temperature (lower = slower), and concentrations of enzyme, substrate, and end product.

Enzyme + substrate → enzyme–substrate complex → enzyme + transformed substrate (product)

Some enzymes require inorganic ions, or **cofactors**. An enzyme complex consisting of the protein portion (**apoenzyme**) and a nonprotein portion (**cofactor**) is called a **holoenzyme**.

Active site—the site on the enzyme at which the substrate binds.

Enzymes (and therefore reactions) can be inhibited in a number of ways:

Competitive inhibition/competitive inhibitors—nonsubstrate molecules compete with substrate molecules for binding at the active site, directly preventing substrate binding.

Noncompetitive inhibition/noncompetitive inhibitors—nonsubstrate molecules bind to a site on the enzyme other than the active site. Binding at this **allosteric site** alters the three-dimensional shape of the enzyme and indirectly prevents substrate binding at the active site.

Feedback inhibition—the end product of the reaction binds at an allosteric site and inhibits substrate binding at the active site.

Genetic regulation—affects enzyme production (see Section 9.12).

4.3   Oxidation and Reduction

When one molecule is oxidized, another is reduced. Such reactions are known as **redox reactions**. It is the **transfer of electrons** that provides energy in a redox reaction.

Oxidation—removal of electrons.

Reduction—gain of electrons.

Cells must have electron donors, electron carriers, and electron acceptors to produce energy.

In certain metabolic reactions, energy is released and captured to form **adenosine triphosphate (ATP)** via the phosphorylation of adenosine diphosphate (ADP).

4.4 Phosphorylation

Phosphorylation—the addition of inorganic phosphate (P_i) to another molecule.

The energy required for phosphorylation may be obtained three ways: oxidative phosphorylation, photosynthetic phosphorylation, and substrate-level phosphorylation.

Oxidative phosphorylation—energy is released as electrons pass through a series of electron acceptors (the electron transport system [see section 4.5.2]) to either oxygen or some other inorganic compound. Coenzymes (e.g., FAD, NAD, NADP) are required.

Photosynthetic phosphorylation—electrons are released as light is absorbed by chlorophyll, and they subsequently pass through the electron transport system (see also section 4.7).

Substrate-level phosphorylation—energy, in the form of a high energy phosphate, is released from a substrate (e.g., a metabolic intermediate) through enzyme activity.

4.5 Carbohydrate Catabolism

The oxidation of glucose (a reduced molecule) provides energy for the cell.

Pathways for the oxidation of organic compounds (e.g., glucose) and the capture of energy in the form of ATP can be divided into three major groups. These groups differ primarily in the substance or molecule that serves as the final electron acceptor.

Fermentation—redox occurs in absence of any added electron acceptor; glucose is only partially broken down.

Aerobic respiration—glucose is completely broken down with molecular oxygen serving as the final electron acceptor.

Anaerobic respiration—an inorganic ion other than molecular oxygen (e.g., NO_3^-, SO_4^{2-}, CO_3^{2-}) serves as the final electron acceptor. Anaerobic respiration yields fewer ATP molecules than does aerobic respiration because only part of the Krebs cycle (see section 4.5.2) is operative under anaerobic conditions.

Glycolysis—the oxidation of glucose to pyruvic acid (and/or glycerol).

Alternatives to glycolysis include the **Entner-Doudoroff pathway** and the **pentose phosphate pathway**. These pathways supply sugars necessary for the synthesis of nucleotides.

4.5.1 Fermentation

Fermentation—oxidative pathways in which *organic* compounds serve as both electron acceptors and electron donors. Energy is released from sugars or other organic molecules in the absence of an added electron acceptor (i.e., an organic molecule serves as the final electron acceptor).

Fermentation produces two ATPs via substrate-level phosphorylation.

There are different kinds of fermentation. Examples include: **Alcoholic fermentation**—which produces ethanol and CO_2. **Heterolactic fermentation**—the pentose phosphate pathway is used to produce lactic acid and ethanol. **Lactic acid fermentation**—pyruvic acid is reduced to lactic acid.

Problem Solving Example:

Does a yeast cell metabolize more efficiently in the presence or in the absence of oxygen? Explain.

A   Under aerobic conditions (in the presence of oxygen), glucose undergoes glycolysis to form pyruvate, which after conversion to acetyl CoA, enters the TCA cycle. The NADH produced is oxidized in the electron transport system provided that oxygen is available as the ultimate electron acceptor. Let us calculate the maximum amount of ATP produced from the oxidation of one mole of glucose. A net gain of two ATP is obtained directly in glycolysis (see figure 4.1). In addition, two NADH are generated. This occurs in the cytoplasm. To enter the electron transport system, the NADH must be transported into the mitochondrion, a process that requires one molecule of ATP for every NADH that crosses the mitochondrial membrane. So instead of three ATP being produced per NADH, each cytoplasmic NADH yields only two net ATP; that is, a total of four ATP are produced by the two NADH generated in glycolysis.

Recall that two 3-carbon molecules are obtained in glycolysis. Each of these two 3-carbon compounds can eventually enter the TCA cycle as acetyl CoA, and later, their electrons enter the electron transport system. Thus, ATP production from the TCA cycle is twice that of a single turn. Two ATP are produced in the cycle via the formation of two molecules of GTP from two molecules of succinyl CoA. The conversion of pyruvate to acetyl CoA yields one NADH while the Krebs cycle yields three NADH. Thus, eight (4 x 2) NADH are produced per glucose at this point. The eight (four per turn) NADH formed in the cycle yield 24 ATP (three ATP per NADH) by means of the electron transport chain. The two $FADH_2$ (one per turn) produce two ATP each via electron transport, giving a total of four ATP. Summation of all the ATP from glycolysis, the citric acid cycle, and electron transport leads to a net production of 36 ATP per mole of glucose fully oxidized.

Under anaerobic conditions, oxygen is not available as a terminal electron acceptor, and the reactions of the electron transport system cease as the reduced intermediates build up. This leads to an accumulation of the TCA cycle intermediates and, since it cannot enter the cycle, an accumulation of pyruvate. Under these conditions in yeast cells, pyruvate is then decarboxylated to acetaldehyde, which is in turn reduced to ethyl alcohol (ethanol), regenerating oxidized NAD^+ for further use in glycolysis. This process is called alcoholic fermentation.

Under anaerobic conditions then, only 2 ATP molecules are produced, as opposed to the 36 ATP produced when glycolysis is supplemented with aerobic respiration. Thus, a yeast cell metabolizes 18 times more efficiently in the presence of oxygen than in the absence of oxygen.

4.5.2 Respiration

Respiration—oxidative pathways in which *inorganic* compounds serve as the final electron acceptors. In aerobic respiration, the final electron acceptor is molecular oxygen; in anaerobic respiration, some other inorganic molecule (e.g., nitrate, carbonate, or sulfate) acts as the final electron acceptor.

In prokaryotes, the complete oxidation of a molecule of glucose yields a total of 38 ATP molecules; in eukaryotes, 36.

The steps or phases of respiration are transition, Krebs cycle, and electron transport.

Transition—following glycolysis, pyruvic acid is converted to acetyl CoA before entering the Krebs cycle. (Some texts do not treat transition separately, but rather consider it as part of the Krebs cycle.)

Krebs cycle (tricarboxylic acid [TCA] cycle, citric acid cycle)—two-carbon groups are oxidized to CO_2 and H_2O. One ATP is produced from each acetyl group. The acetyl CoA is processed so that the hydrogen atoms can be transferred to the electron transport system and oxidized for energy.

Electron transport system (electron transport chain)—a sequence of redox reactions that provide energy for the oxidative phosphorylation of three molecules of ADP to ATP. This process involves the coenzymes nicotinamide adenine dinucleotide (NAD), nicotinamide adenine dinucleotide phosphate (NADP), and flavin adenine dinucleotide (FAD), as well as carrier molecules such as flavoproteins, cytochromes, and quinones (coenzyme Q). In eukaryotes, oxidative phosphorylation takes place along the cristae of the mitochondria.

Theory of chemosmosis—as electrons move through a series of carriers or acceptors, protons being pumped across a membrane (the inner mitochondrial membrane in eukaryotes or the plasma membrane in prokaryotes) generate a **proton motive force**, the energy of which is sufficient to produce ATP.

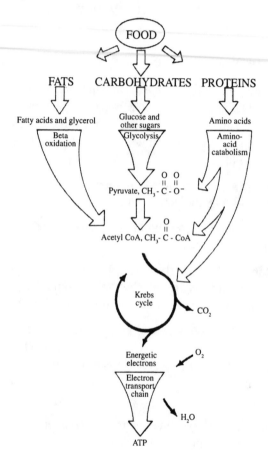

Figure 4.1 Summary: Catabolic Pathways

4.6   Lipid and Protein Catabolism

Lipids are hydrolyzed into glycerol and fatty acids. Glycerol is then catabolized in glycolysis; fatty acids undergo beta oxidation to acetyl CoA, which enters the Krebs cycle.

Proteins are first broken down to amino acids, which must then be transaminated, decarboxylated, or dehydrogenated before their subsequent catabolism in either glycolysis, the Krebs cycle, or fermentation.

4.7 Photosynthesis

Photosynthesis—the utilization of light energy from the sun to synthesize carbohydrates from CO_2. The process involves conversion of the light energy into chemical energy that can be used for carbon fixation. There are two different sets of reactions in photosynthesis—the light reactions and the dark reactions (also known as the Calvin-Benson cycle).

In the **light reactions**, electrons from chlorophyll travel through an electron transport chain, producing ATP via chemosmosis. The light reactions can involve **cyclic photophosphorylation,** wherein the electrons return to chlorophyll, or **photolysis accompanied by noncyclic photophosphorylation**, wherein the electrons are used to reduce NADP; electrons from H_2O or H_2S return to chlorophyll.

In the **dark reactions**, CO_2 is reduced to synthesize carbohydrates.

Problem Solving Example:

 Why are autotrophic organisms necessary for the continuance of life on earth? Do all autotrophs require sunlight?

Autotrophic organisms have the capacity to generate all needed energy from inorganic sources. Heterotrophic organisms can only utilize the chemical energy present in organic compounds. There are two main types of autotrophs—photosynthetic organisms and chemosynthetic organisms. Photosynthetic autotrophs obtain energy from sunlight, and convert the radiant energy of sunlight to the chemical energy stored in the bonds of their organic compounds. Green plants obtain CO_2 from the atmosphere and minerals and water from the soil. Algae and photosynthetic bacteria absorb dissolved CO_2, water, and minerals through their cell membranes. Using energy from sunlight, the photosynthetic autotrophic organism converts CO_2, water, and minerals into all the constituents of the organism. Chemosynthetic organisms are much less common than photosynthetic organisms, and are always bacteria. Chemosynthetic bacteria do not require sunlight, and obtain energy by oxidizing certain substances. Two examples are the nitrifying bacteria, which oxidize ammonia to nitrate (NO_2^-) or nitrates (NO_3^-) ultimately, and the sulfate bacteria, which oxidize sulfur to sulfates. The energy released from these chemical reactions is converted to a form of chemical energy utilized by the organism.

All the organisms that carry on respiration, that is, oxidize organic compounds to carbon dioxide and water, require oxygen. Respiration is the process by which heterotrophs obtain energy. Chemosynthetic bacteria also require oxygen in order to carry out oxidation of inorganic substances. The only source of oxygen on earth is the photosynthetic autotrophs. These organisms convert CO_2 and water to organic compounds, utilizing sunlight to provide energy, and generate O_2 in the process. If there were no green plants or photosynthetic marine organisms, the oxygen present in the atmosphere would quickly be used up by animals, bacteria, and fungi.

The autotrophs are also responsible for providing the heterotrophs with organic nourishment. Sunlight is the most important source of energy on the earth, and it is only the photosynthetic autotrophs that can utilize this energy, converting it to chemical energy in organic compounds. Heterotrophs utilize the organic compounds produced by the autotrophs. Heterotrophs that obtain organic nourishment from other heterotrophs are also ultimately dependent upon autotrophs for nourishment, because the animals that are their prey have either directly or indirectly (through another animal) utilized the organic material of plants or algae. Photosynthetic autotrophs provide the earth with an energy source for living organisms and with oxygen. If photosynthetic autotrophs were not present, all life on earth would eventually cease as the food and O_2 would become depleted.

4.8 Other Anabolic Pathways

Lipids are synthesized from glycerol (derived from dihydroxyac-etone phosphate) and fatty acids (derived from acetyl CoA).

Proteins are synthesized from amino acids that are synthesized from intermediates in the Krebs cycle or other phases of carbohydrate metabolism.

4.9 Nutritional Modes

Autotrophs are organisms that can synthesize all of their essential biochemical compounds using *inorganic* carbon (CO_2) as their sole carbon source. All autotrophs use the same pathway for carbon fixation, i.e., the Calvin-Benson cycle (the "dark reactions" of photosynthesis). Autotrophs that use sunlight as their energy source are referred to as photosynthetic autotrophs or **photoautotrophs**; those that use inorganic compounds (e.g., hydrogen or hydrogen sulfide) as their energy source are known as chemosynthetic autotrophs or **chemoautotrophs**.

Heterotrophs are organisms that lack the ability to fix CO_2. They must obtain their carbon from an *organic* source. Some heterotrophs use light as an energy source and are known as photosynthetic heterotrophs, or **photoheterotrophs**. Chemosynthetic heterotrophs, or **chemoheterotrophs**, use organic compounds for their energy source as well as for their carbon source. Chemoheterotrophs carry out the biochemical reactions known as glycolysis, fermentation, and respiration to synthesize carbohydrates and other biochemical compounds and macromolecules.

Photosynthetic organisms oxidize water to produce oxygen and are therefore referred to as **oxygenic phototrophs**. Examples of oxygenic phototrophs are **cyanobacteria** and **algae**.

Organisms that cannot undergo photosynthesis in the presence of oxygen are **anoxygenic phototrophs**. That is, photosynthesis in these organisms does not produce oxygen. Examples of anoxygenic phototrophs are **green sulfur** and **purple sulfur bacteria**.

Table 4.1 Nutritional Modes of Organisms

	Energy Source	Carbon Source
Photoautotrophs	Light	Inorganic
Chemoautotrophs	Inorganic chemical reactions	Inorganic
Photoheterotrophs	Light	Organic
Chemoheterotrophs	Organic chemical reactions	Organic

Problem Solving Examples:

Q What basic nutritional requirements do all living organisms have in common? Compare phototrophs and chemotrophs, autotrophs and heterotrophs.

A Organisms, ranging from bacteria to humans, share a set of nutritional requirements necessary for normal growth. These requirements must be known in order to cultivate microorganisms in the laboratory, and pure cultures may be obtained through the preparation of appropriate selective growth media.

All organisms require a source of energy. Green plants and some bacteria can utilize radiant energy (from the sun) and are thus called phototrophs. Animals and nonphotosynthetic bacteria must rely on the oxidation of chemical compounds for energy and are thus called chemotrophs.

All organisms require a carbon source. Plants and many bacteria require only carbon dioxide as their carbon source. They are termed autotrophs. Animals and other bacteria require a more reduced form of carbon, such as an organic carbon compound. Sugars and other carbohydrates are examples of organic carbon compounds. Organisms that have this requirement are termed heterotrophs. They depend upon autotrophs for their organic form of carbon, which they use as both a carbon source and an energy source.

All organisms require a nitrogen source. Plants utilize nitrogen in the form of inorganic salts such as potassium nitrate (KNO_3), while animals must rely on organic nitrogen-containing compounds such as amino acids. Most bacteria utilize nitrogen in either of the above forms, although some bacteria can use atmospheric nitrogen.

All organisms require sulfur and phosphorus. While phosphorus is usually supplied by phosphates, sulfur may be supplied by organic compounds, by inorganic compounds, or by elementary sulfur.

All organisms need certain metallic elements, and many require vitamins. The metallic elements include sodium, potassium, calcium, magnesium, manganese, iron, zinc, and copper. While vitamins must be furnished to animals and to some bacteria, there are certain bacteria capable of synthesizing the vitamins from other nutrient compounds.

Finally, all organisms require water for growth. For bacteria and plants, all the above nutrients must be in solution in order to enter the organism.

Bacteria show considerable variation in the specific nutrients required for growth. For example, all heterotrophic bacteria require an organic form of carbon, but they differ in the kinds of organic compounds they can utilize. Different bacteria utilize nitrogen in its various forms. Some require several kinds of amino acids and vitamins, while others require only inorganic elements.

Q Most bacteria are heterotrophic, requiring an organic form of carbon, such as glucose, which they oxidize to obtain energy. How do chemotrophic bacteria that can utilize carbon dioxide as a sole carbon source acquire energy?

A Autotrophs are organisms that require only carbon dioxide as a carbon source, from which they can construct the carbon skeletons of all their organic biomolecules. The carbon dioxide is reduced to glyceraldehyde-3-phosphate in order to synthesize carbohydrates. This reductive process requires much energy. Since glucose is not available to be oxidized for energy, these autotrophic bacteria must oxidize inorganic compounds. They are called chemoautotrophs, since they obtain their energy by oxidizing chemical compounds (as opposed to photoautotrophs, which obtain their energy from light). The inorganic compounds oxidized by the various bacteria include molecular hydrogen, ammonia, nitrite, sulfur (sulfide ions), and iron (ferrous ions). These oxidations result in electrons that enter the respiratory chain, with the concomitant production of ATP. ATP is then used as a source of energy in the reduction of CO_2.

Bacteria of the *Hydrogenomonas* genus obtain energy through the oxidation of hydrogen gas. They possess an enzyme, hydrogenase, which catalyzes the following reaction:

$$H_2 + 1/2\ O_2 \longrightarrow H_2O + 2e^-$$

The electrons are transferred to NAD to form $NADH_2$, which is then oxidized in the respiratory chain, yielding ATP.

The bacterium *Nitrosomonas* obtains energy by the oxidation of the ammonium ion:

$$2NH_4^+ + 3O_2 \longrightarrow 2NO_2^- + 2H_2O + 4H^+ + 2e^-$$

The electrons produced enter the respiratory chain, where they are passed down to O_2, forming ATP along the chain.

Bacteria of the *Nitrobacter* genus obtain their energy through the oxidation of nitrite ions into nitrate ions:

$$2NO_2^- + O_2 \longrightarrow 2NO_3^- + 2e^-$$

Again, the electrons produced are used in the formation of ATP.

The oxidation of ammonia into nitrate is called nitrification. It is one of the most important activities of autotrophic bacteria since it provides the form of nitrogen most available to plants. Nitrification is carried out in two stages, with the first stage involving *Nitrosomonas* and the second stage featuring the oxidation of nitrites by *Nitrobacter*. The opposite process, called nitrate respiration, is the reduction of nitrate to ammonia. This process is carried out by several heterotrophic bacteria under anaerobic conditions. The oxygen of the nitrate serves as the hydrogen acceptor (under aerobic conditions, molecular oxygen would normally serve as the final electron or hydrogen acceptor). The overall reaction is as follows:

$$HNO_3 + 4H_2 \longrightarrow NH_3 + 3H_2O$$

Nitrification should not be confused with nitrogen fixation or denitrification. Nitrogen-fixing microorganisms in the soil use molecular nitrogen in the atmosphere as their source of nitrogen and convert it into ammonia. The ammonia is then used in the synthesis of proteins and other nitrogenous substances. Denitrification is the reduction of nitrates to molecular nitrogen carried out by certain bacteria, such as *Pseudomonas*.

CHAPTER 5

Transport of Molecules

5.1 Transport

Transport, or the movement of materials across the plasma membrane (or any cellular membrane), may or may not require energy expenditure from the cell. **Passive processes** (e.g., diffusion and osmosis) are those in which materials move along a concentration gradient from an area of higher concentration to an area of lower concentration and do not require energy. Conversely, **active transport** and **group translocation** involve movement of materials against a concentration gradient (i.e., from an area of lower concentration to an area of higher concentration), and they do require energy.

5.2 Simple Diffusion

Simple diffusion—materials, generally small molecules and ions, move from an area of higher concentration to an area of lower concentration until a state of equilibrium is reached (i.e., the concentration is the same on both sides of the membrane).

5.3 Osmosis

Osmosis refers specifically to the movement of water molecules across a selectively permeable membrane, from an area of higher (water) concentration to an area of lower (water) concentration until equilibrium is reached.

5.3.1 Hypotonic, Isotonic, and Hypertonic

Hypotonic environment—the concentration of solute in the extra-cellular medium is lower than the concentration of solute within the cell. A hypotonic environment forces the cell to expend energy via active transport or group translocation in order to move molecules into the cell against the concentration gradient. There is a danger that the cell will rupture due to the tendency of osmosis to move water into the cell under these conditions.

Isotonic environment—the concentration of solute in the extracellular medium is equal to the concentration of solute within the cell.

Hypertonic environment—the concentration of solute in the extracellular medium is greater than the concentration of solute within the cell. The cell can use diffusion processes to move molecules along the concentration gradient into the cell. Water will tend to move out of the cell by osmosis (**plasmolysis**), resulting in cell shrinkage (**crenation**).

Problem Solving Example:

Q A typical bacterial cell has a cell wall and plasma membrane. If the cell wall is removed by treatment with lysozyme, an enzyme that selectively dissolves the cell wall material, will the bacteria live? Explain. What is the chemical difference between the cell walls of bacteria and plants?

A The main function of the bacterial cell wall is to provide a rigid framework or casing that supports and protects the bacterial cell from osmotic disruption. Most bacteria live in a medium that is hypotonic relative to the bacterial protoplasm: i.e., the bacterial protoplasm is more concentrated than the medium, and water tends to enter the cell. If intact, the cell wall provides a rigid casing, which prevents the cell from swelling and bursting. If the cell wall were destroyed, water would enter the cell and cause osmotic lysis (bursting). If the bacterium is placed in a medium that has the same osmotic pressure as the bacterial cell contents, osmotic disruption would not occur when the cell wall was dissolved. A bacterium devoid of its cell wall is called a protoplast. Protoplasts can live in an isotonic medium, (medium having equal osmotic pressure as the bacterial protoplast). If the bacterium was placed in a hypertonic medium, water would leave the cell and plasmolysis (shrinkage) would ensue.

The cell wall varies in thickness from 100 to 250 Å (an angstrom, Å, is 10^{-10} meter) and may account for as much as 40% of the dry weight of the cell. While the cell wall in eukaryotes is composed of cellulose, in bacteria, the cell wall is composed of insoluble peptidoglycan. Peptidoglycan consists of sugars (N-acetylglucosamine and N-acetylmuramic acid) and amino acids, including diaminopinelic acid, an amino acid unique to bacteria. In gram-negative bacteria, the peptidoglycan constitutes a much smaller fraction of the wall component than it does in gram-positive bacteria. The higher lipid content in the cell walls of gram-negative bacteria accounts for the differences in gram-staining.

The plasma membrane, a thin covering immediately beneath the cell wall, is too fragile to provide the support needed by the cell. Instead, this semipermeable membrane controls the passage of nutrients and waste products into and out of the cell.

5.4 Facilitated Diffusion

Facilitated diffusion—movement is from an area of higher concentration to one of lower concentration, but substances being transported across the membrane first combine with membrane **carrier proteins (permeases)**.

5.5 Active Transport

Active transport—**permeases** transport molecules across the membrane against the concentration gradient from an area of low concentration to an area of higher concentration. Energy (ATP) is required.

5.6   Group Translocation

Group translocation—a process unique to prokaryotes, in which energy is used to chemically alter the substance being transported. Once within the cell, the modified substance can accumulate within the cell. This process requires a high-energy phosphate compound such as ATP or PEP (phosphoenolpyruvic acid) for energy.

5.7   Endocytosis and Exocytosis

Unique to eukaryotic cells, **endocytosis** and **exocytosis** are two additional transport processes. They are important in moving large quantities of a substance. Both involve the formation of vesicles from the plasma membrane. Endocytosis is the formation of a vesicle that moves into the cell (e.g., by **phagocytosis**). Exocytosis involves the formation of vesicles that are moved out of the cell (e.g., by secretion).

Quiz: Microbial Metabolism & Transport of Molecules

1.   The majority of ATP molecules derived from nutrient metabolism are generated by (the)

   (A)   anaerobic fermentation and glycolysis.

   (B)   fermentation and electron transport chain.

   (C)   glycolysis and substrate phosphorylation.

   (D)   Krebs cycle and electron transport chain.

   (E)   substrate phosphorylation.

2.   Bacteria that can effectively carry out metabolism in the presence or absence of oxygen are described as

   (A)   aerobic.

   (B)   anaerobic.

    (C)   facultative anaerobes.

    (D)   fermentative microbes.

    (E)   glycolytic.

3.   In comparing glycolysis under aerobic and anaerobic conditions,

    (A)   pyruvate is converted to lactate under aerobic conditions, while it's converted to acetyl CoA under anaerobic conditions.

    (B)   pyruvate is converted to acetyl CoA under aerobic conditions, while it's converted to lactate under anaerobic conditions.

    (C)   oxygen is used in anaerobic glycolysis, while it is not used in aerobic conditions.

    (D)   humans are only capable of aerobic metabolism.

    (E)   lactate is converted to pyruvate under aerobic conditions, while the reverse occurs under anaerobic conditions.

4.   Fermentation

    (A)   results in the formation of lactic acid.

    (B)   does not require oxygen.

    (C)   does require oxygen.

    (D)   produces large amounts of energy.

    (E)   occurs only in bacteria.

5.   A protozoan cell is immersed in pure water. Select the correct statement from the following about the cell's behavior and environment.

    (A)   The cell will gain water.

    (B)   The cell will lose water.

(C) The exterior environment is hypertonic.

(D) The cell's interior is hypotonic.

(E) The cell is isotonic to the outside.

6. This process accounts for the ability of freshwater alga, *Nitella*, to accumulate a concentration of potassium ions more than a thousand times greater than that of the surrounding water.

(A) Active transport

(B) Osmosis

(C) Phagocytosis

(D) Simple diffusion

(E) Facilitated diffusion

7. The opposite of feedback inhibition is

(A) positive feedback, in which an increase in end product increases enzyme activity.

(B) negative feedback, in which an increase in end product decreases enzyme activity.

(C) competitive inhibition, in which a molecule similar in structure to the substrate blocks the active site of the enzyme.

(D) positive modulation, in which a decrease in substrate increases enzyme activity.

(E) end-product inhibition, in which the cell is stopped from wasting chemical resources.

8. Marine amoebas lack a contractile vacuole. This is because

(A) they have alternative means of osmoregulation.

(B) their internal environment is isotonic to their surroundings.

   (C) their internal environment is hypotonic to their surroundings.

   (D) during the course of evolution, they have lost the need for a contractile vacuole.

   (E) their internal environment is hypertonic to their surroundings.

9.   A vacuole containing material to be expelled travels to the cell membrane and fuses with it. After fusion has been completed, the site of contact opens up and the contents of the vacuole are jettisoned out of the cell. This process is known as

   (A) endocytosis.

   (B) phagocytosis.

   (C) pinocytosis.

   (D) exocytosis.

   (E) None of the above.

10.   A bacterium that can grow on the most minimal medium and that synthesizes all the essential organic compounds it needs is a(n)

   (A) auxotroph.

   (B) heterotroph.

   (C) organotroph.

   (D) chemio organotroph.

   (E) prototroph.

ANSWER KEY

1.	(D)	6.	(A)
2.	(C)	7.	(A)
3.	(B)	8.	(B)
4.	(B)	9.	(D)
5.	(A)	10.	(E)

CHAPTER 6

Bacterial Growth

6.1 Growth of Bacterial Populations

Growth is the orderly increase in quantity of all cellular components and structures. The growth of an individual cell leads to an increase in size and is generally followed by cell division.

Vegetative cell—one that is actively growing and dividing.

Cell division in bacteria usually occurs by **binary fission**, in which the cell divides into two new (approximately equal and identical) cells.

Cell division by **budding**, in which the new cell develops as a small outgrowth from the surface of the existing (parent) cell, occurs in some bacteria and in yeast.

Other bacteria may reproduce by fragmentation or by aerial spore formation. Although it does occur under favorable conditions as well, spore formation (**sporulation**) generally serves to allow the organism to withstand long periods of unfavorable conditions such as extreme temperatures or dryness.

Microbial growth is assayed as an increase in cell number or mass of a **population** of cells.

Generation time (doubling time)—the time it takes for an individual cell to divide or for a population of cells to double. Bacterial growth follows a **logarithmic** (exponential) progression, e.g., 2 cells → 4 cells → 8 cells → 16 cells, etc.

$$\text{Generation time} = \frac{t}{3.3 \; \log \dfrac{B_0}{B_1}}$$

where t is the time interval between cell number measurements B_0 and B_1 during log phase growth; B_0 is the initial number of bacteria; and B_1 is the number of bacteria after time t.

When bacteria are placed in fresh, nutrient-rich medium, they exhibit four characteristic **phases of population growth**: lag phase, log phase, stationary phase, and death phase (see Figure 6.1).

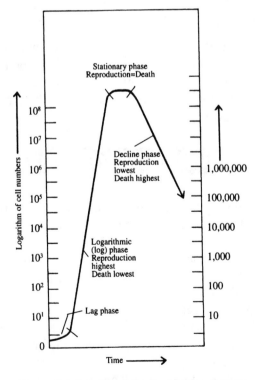

Figure 6.1 Growth of Bacterial Populations

Lag phase—cells are metabolically active but are not dividing. This is a period when the cells are resynthesizing enzymes, coenzymes, etc., necessary for growth and division.

Log phase—bacteria are growing and dividing at an exponential, or logarithmic, rate. This is the period of fastest growth; the generation time is maximal and constant. All nutrients and molecules needed for growth are in good supply.

Stationary phase—at this point, the medium is becoming depleted in some nutrients, and toxic quantities of waste materials may be accumulating. The number of new cells produced is offset by the number of cells that are dying; thus, the total number of viable cells remains approximately constant.

Death phase—conditions are becoming less and less conducive to cell growth. Cells are dying more rapidly than new ones are being formed, resulting in a logarithmic decrease in the number of cells.

Continuous culture—when fresh medium is continuously provided to a population of cells in log phase, the population continues to grow. In this way, a continuous culture can be maintained over long periods of time. A **chemostat** is a device that allows for the continuous addition of fresh medium.

Problem Solving Examples:

 How does the term "growth" as used in bacteriology differ from the same term as applied to higher plants and animals?

When a small number of bacteria are transferred into the proper medium and incubated under the appropriate physical conditions, a tremendous increase in the number of bacteria results in a short time. As applied to bacteria and microorganisms, the term "growth" refers to an increase in the entire population of cells. When we speak of the growth of plants and animals, we usually refer to the increase in size of the individual organism. The growth of bacteria involves the increase in numbers of cells over the initial quantity used to start the culture (called the inoculum). Some species of bacteria require only a day to reach their maximum population size, while others require a longer period of incubation. Growth can usually be determined by measuring cell number, cell mass, or cell activity.

 Describe the most common process of bacterial reproduction.

The most important process in the growth of bacterial populations is binary (transverse) fission. This is a type of cell division in which two identical daughter cells are produced as a result of the division of the parent cell. New cell wall material begins to form on the inner surface of the wall of the parent cell at a point midway along its length and this new wall material invaginates, dividing the cellular material of the parent cell evenly into two halves. The cell is then separated in two by the completion of the transverse wall. Each daughter cell possesses a complete set of genetic information. The genes, the units of inheritance, are arranged in sequence along a single, circular chromosome composed of DNA. During reproduction, the DNA is replicated and the two chromosomes move apart into separate nuclear areas in each half of the parent cell before the completion of the transverse cell wall. Each daughter cell has genetic information identical to the parent cell. This process of giving rise to new individuals by cell division is termed asexual reproduction.

Although binary fission is the major method of bacterial reproduction, another form of asexual reproduction, budding, is observed in some bacteria. Budding involves an outgrowth of the parental cell, which enlarges and separates to form a new cell.

 A biology student places one bacterium in a suitable medium and incubates it under appropriate conditions. After 3 hours and 18 minutes, he determines that the number of bacteria now present is 1,000. What is the generation time of this bacteria?

Bacterial growth usually occurs by means of binary fission with one cell dividing into two and these two cells dividing into four, etc. The population at any time can be represented by a geometric progression (1, 2, 4, 8, …). The time interval required for each division (or for the entire population to double) is called the generation time. Different bacteria have different generation times, ranging from 20 minutes for *Escherichia coli* to 33 hours for *Treponema pallidum* (the bacterium responsible for syphilis). The generation time also differs for the same bacterial species under different environmental conditions.

To determine the generation time (G) of a bacterial population, we must know the number of bacteria present initially (N_0), the number of bacteria present at the end of a given time period (N), and the time period (t). We can determine the generation time by using some simple mathematical expressions. If we start with a single bacterium, the total population (N) after the nth generation is 2^n:

$$N = 1 \times 2^n$$

For example, at the end of three generations, we would have eight bacteria ($2^3 = 8$). However, since we usually start with many bacteria, we must modify the formula to account for more than one parental bacterium:

$$N = N_0 \times 2^n.$$

Solving for n, the number of generations, we get

$$\log N = \log N_0 + n \log 2.$$

By substituting .301 for the log 2 and rearranging,

$$n = \frac{\log N - \log N_0}{0.301} = 3.3\,(\log N - \log N_0).$$

Since the difference of the logarithms of two numbers is the logarithm of the quotient of the two numbers,

$$N = 3.3\ \log\ \frac{N}{N_0}.$$

The generation time G is simply equal to the time elapsed between N_0 and N divided by the number of generations:

$$G = \frac{t}{n} = \frac{t}{3.3 \log \frac{N}{N_0}} .$$

In our example: $N = 1,000$, $N_0 = 1$, and $t = 198$ minutes. Substituting into the general formula,

$$G = \frac{198}{3.3 \times \log 1,000}$$

$$G = \frac{198}{3.3 \times 3}$$

$$= 20 \text{ minutes}$$

The time interval required for this particular bacterial species to divide is 20 minutes.

 Discuss the factors for the existence of each phase of the bacterial populations.

A If one inoculates a flask of nutrient broth with a given number of bacteria and follows the rate of growth during the incubation period, there is found a series of different growth rates. A plot of the logarithms of the number of cells versus time is used to illustrate the growth curve.

There is an initial period of no growth, followed by one of rapid growth, and then there is a leveling off. The last period is one of decline in the population. The curved portions (see figure 6.1) designate the transitional period between phases. These represent periods where the bacteria enter a new phase.

Following the addition of inoculum into the new medium, the bacterial population does not increase, but each individual bacterium increases in size. During this period, called the lag phase, the bacteria are physiologically active and are adapting to the new environment, but the bacterial population number remains constant. In addition, the bacteria may alter the environment for their means (e.g., excretion of CO_2 to lower pH, so that more favorable growth conditions are achieved). At the end of the lag phase, the bacteria start to divide.

The lag phase is followed by the logarithmic or exponential phase where the cells divide at a constant and maximal rate according to their generation time. Most of the bacteria during this phase are uniform in terms of metabolic activity, unlike during the other phases.

After the log phase, growth begins to level off in the stationary phase. The population remains constant because of a cessation of reproduction or an equalization of growth and death rates. The decreased growth is usually due to the exhaustion of nutrients or the production of toxic or inhibitory products.

The final phase is called the death phase. Here, the bacteria die faster than they are being produced, if any reproduction is occurring at all. Depletion of nutrients and accumulation of inhibitory products, such as acid, cause the increased death rate. The number of bacteria increases exponentially in the log phase, and the number of bacteria decreases exponentially in the death phase. Some species die rapidly, so that few living cells remain after about three days, while other species die after several months.

6.2 Ways to Measure Growth of Bacterial Populations

Growth can be measured either directly or indirectly. **Direct methods** include direct microscope counts, standard plate counts, filtration, and the most probable number technique. **Indirect methods** include measurement of metabolic activity (e.g., oxygen consumption), dry weight, or turbidity (cloudiness of a suspension).

Viable count—a count that does *not* include dead cells. Standard plate counts, filtration, and most probable number are all viable count techniques. Measurement of metabolic activity, although an indirect method, also measures only viable cells. Other methods of indirect measure, as well as direct microscope counts, fail to distinguish between living and dead cells; thus, all cells, both living and dead, are included in the count.

Direct microscope counts—a measured volume of a bacterial suspension is placed on a special type of microscope slide known as a Petroff-Hauser bacterial counter (hemocytometer) and cells are counted under the microscope.

Standard plate counts are the most commonly used methods. They are based on the premise that a viable cell, when placed on the appropriate medium, will give rise to a bacterial colony; thus, cells are measured as **colony forming units** (CFUs). The formula for computing CFU is:

$$CFU = \frac{\text{number of colonies}}{\text{volume of sample}} \times \text{reciprocal of the dilution}$$

There are two standard plate count methods: the **spread plate**, in which a small sample is spread over the surface of an agar plate containing the appropriate medium, and the **pour plate**, in which the sample is mixed in with the melted agar medium before the plate is poured. Proper dilution of the sample is required as only those plates with between 30 and 300 colonies are considered valid for counting.

Filtration—a bacterial suspension, or dilution thereof, is filtered through a membrane. Cells that are retained on the membrane are resuspended and transferred to an agar plate for growth and subsequent counting of colonies.

Most probable number (MPN)—this method is based on a statistical estimation for the number of cells. It involves dilution of the sample to an estimated concentration of less than one cell per milliliter. One milliliter aliquots of the dilution are used to inoculate test tubes of fresh liquid medium, which are then observed for growth. If growth occurs, the dilution contained at least one cell. It is important to dilute the samples properly because if the sample is overly diluted, there is no growth and the count is invalid; conversely, if the sample is not diluted enough, all tubes show growth and again the count is invalid.

Problem Solving Example:

 What are some methods of measuring bacterial growth?

There are several ways of measuring the growth of microbial populations. Direct measurements include: plate counts, serial dilutions, the pour plate/spread plate method, filtration, MPN method, and direct microscopic count. Indirect measurements include: turbidity, metabolic activity, and dry weight.

The standard plate count assumes that each viable bacterium grows into a single colony on an agar plate by either the pour plate or spread plate method. Serial dilutions approximate the number of cells in one ml of the original sample by multiplying the number of viable cells times the reciprocal of the dilution. In the filtration method, a filter membrane is used to transfer viable bacteria to a growth medium. MPN (most probable number) is a statistical estimation. Direct microscopic count depends on the use of a hemocytometer.

Turbidity is measured by a spectrophotometer. Acid production and oxygen consumption are examples of measurement of metabolic activity. Dry weight is used as an indirect measurement of growth and is especially useful for fungi.

6.3   Physical and Chemical Requirements for Growth

For any microbe, there is a range of temperatures and pH within which it will grow. Conditions beyond those ranges may inhibit growth or even kill the organism.

Temperature—microbes vary in their range of temperature tolerance. Organisms with a narrow temperature range are said to be **stenothermal.** Those with a wide temperature range are **eurythermal.**

Psychrophiles (cryophiles) are organisms with low temperature optima, generally between 0°C and 20°C.

Mesophiles prefer mid-range temperatures, 20°C – 40°C.

Thermophiles have high temperature optima, >40°C.

Cardinal temperatures (minimum, optimum, and maximum growth temperatures) define the growth range of an organism:

Minimum growth temperature—the organism cannot grow below this temperature.

Optimum growth temperature—temperature at which it grows best/most rapidly.

Maximum growth temperature—temperature above which growth does not occur.

The **temperature range** of an organism extends from the minimum growth temperature through the maximum growth temperature.

pH—a measure of acidity or alkalinity. pH values between 5 and 9 are common in nature, and the pH optima of many organisms falls within this range. However, there are organisms that can live at a pH of as low as 0 or as high as 12. Organisms preferring low pH environments are called **acidophiles.** Generally, an organism cannot tolerate drastic changes in pH; the range of acceptable pH values is usually only about one pH unit.

Chemically, bacteria require nitrogen, carbon, and energy sources; water; vitamins and growth factors; and essential mineral salts (e.g., phosphorus, potassium, iron, sulfur).

Nitrogen (organic or inorganic)—necessary for synthesis of proteins and nucleic acids. NH_4^+, NO_3^-, and protein decomposition serve as nitrogen sources. Some bacteria are able to fix inorganic nitrogen (N_2).

Carbon—every organism must have a carbon source—autotrophic organisms can fix inorganic carbon (CO_2); heterotrophic organisms require an organic source of carbon (e.g., glucose).

Water—most metabolically active bacteria need a watery environment in which to live. If the salt concentration of the surrounding water is too high (i.e., is hypertonic), crenation of the cell may occur;

however, some organisms (**halophiles**) require moderate to high salt concentrations.

Oxygen—microorganisms vary in their oxygen requirements. **Obligate aerobes** are those organisms that must have relatively large amounts of oxygen to grow. **Facultative anaerobes** can metabolize aerobically when oxygen is available, or anaerobically when it is not. Oxygen must be present for **microaerophiles** to grow, but only at low pressures (<0.2 atm O_2). Oxygen is toxic to **obligate anaerobes**, which cannot grow in the presence of oxygen. **Aerotolerant** organisms cannot metabolize aerobically, but are not harmed by the presence of oxygen.

Problem Solving Examples:

Besides temperature, what other physical conditions must be taken into account for the growth of bacteria?

Although all organisms require small amounts of carbon dioxide, most require different levels of oxygen. Bacteria are divided into four groups according to their need for gaseous oxygen.

Aerobic bacteria can only grow in the presence of atmospheric oxygen. *Shigella dysenteriae* are pathogenic bacteria (causing dysentery) that require the presence of oxygen.

Anaerobic bacteria grow in the absence of oxygen. Obligate anaerobes grow only in environments lacking O_2. *Clostridium tetani* are able to grow in a deep puncture wound since air does not reach them. These bacteria produce a toxin that causes the painful symptoms of tetanus (a neuromuscular disease).

Facultative anaerobic bacteria can grow in either the presence or absence of oxygen. *Staphylococcus*, a genera commonly causing food poisoning, is a facultative anaerobe.

Microaerophilic bacteria grow only in the presence of minute quantities of oxygen. *Propionibacterium*, a genus of bacteria used in the production of Swiss cheese, is a microaerophile.

The growth of bacteria is also dependent on the acidity or alkalinity of the medium. For most bacteria, the optimum pH for growth lies between 6.5 and 7.5, although the pH range for growth extends from pH 4 to pH 9. Some exceptions do exist, such as the sulfur-oxidizing bacteria: *Thiobacillus thiooxidans* grow well at pH 1. Often the pH of the medium will change as a result of the accumulation of metabolic products. The resulting acidity or alkalinity can inhibit further growth of the organism or can actually kill the organism. This phenomenon can be prevented by addition of a buffer to the original medium. Buffers are compounds that act to resist changes in pH. During the industrial production of lactic acid from whey by *Lactobacillus bulgaria* lime, Ca (OH)$_2$ is periodically added to neutralize the acid. Otherwise, the accumulation of acid would retard fermentation.

 How does a facultative anaerobe differ from an obligate anaerobe?

Organisms that can live anaerobically are divided into two groups. The obligate, or strict, anaerobe cannot use oxygen and dies in the presence of oxygen. These include denitrifying bacteria of the soil, which are responsible for reducing nitrate to nitrogen, and methane-forming bacteria, which produce marsh gas. Some obligate anaerobes are pathogenic to man; these include *Clostridium botulinum*, responsible for botulism, a fatal form of food poisoning; *Clostridium perfringens*, which causes gas gangrene in wound infections; and *Clostridium tetani*, which causes the disease tetanus.

The facultative anaerobes can live either in the presence or absence of oxygen. Under anaerobic conditions, they obtain energy from a fermentation process; under aerobic conditions, they continue to degrade their energy source anaerobically (via glycolysis) and then oxidize the products of fermentation using oxygen as the final electron acceptor. Yeast will grow rapidly under aerobic conditions but will still continue to live when oxygen is removed. It reproduces more slowly but maintains itself by fermentation. Winemakers take advantage of this behavior by first aerating crushed grapes to allow the yeasts to grow rapidly. They then let the mixture stand in closed vats while the yeasts convert the grape sugar anaerobically to ethanol.

Quiz: Bacterial Growth

1. Heating a test tube culture full of bacteria, thus killing them all, is

    (A)   a density-dependent factor.

    (B)   an intrinsic factor.

    (C)   a result of exponential growth.

    (D)   a result of predation.

    (E)   None of the above.

2. One bacterial cell is placed into a nutrient broth in a test tube at noon. Its generation time is 20 minutes. By 2:00 p.m., the size of the population of bacteria in the test tube is

    (A)   2.

    (B)   16.

    (C)   32.

    (D)   64.

    (E)   128.

3. A correct statement about the population's stationary phase is

    (A)   cells are not dying.

    (B)   cells are not reproducing.

    (C)   cell production rate equals cell death rate.

    (D)   cell death rate exceeds cell production rate.

    (E)   cells reproduce too rapidly.

4. Bacteria may be classified into physiological groups according to the range of temperatures that will permit their growth. The type most suited for cold conditions are the

    (A) mesophiles.

    (B) psychrophiles.

    (C) thermophiles.

    (D) thermophobes.

    (E) poikilotherms.

5. In cell culture, the terms "synchronous growth" and "balanced growth" are used. It can therefore be concluded that

    (A) they are synonymous.

    (B) balanced growth refers to a situation where all routine metabolic and reproductive functions are at a maximum in each cell, so that increases in DNA synthesis are proportional to increases in protein synthesis which are proportional to increases in RNA synthesis.

    (C) balanced growth defines a situation where all cells are at the same point in the cell cycle at the same time.

    (D) the description in (B) would be correct if it started with the words "synchronous growth."

    (E) None of the above.

6. After incubation of two bacteria cultures for 132 minutes, the number of bacteria present is 20,000. What is the generation time?

    (A) 2 minutes

    (B) 5 minutes

    (C) 10 minutes

    (D) 30 minutes

    (E) Greater than 30 minutes

7. The point at which the population of *E. coli* has reached its optimal yield is the

   (A) stationary phase.

   (B) lag phase.

   (C) logarithmic phase.

   (D) death phase.

   (E) None of the above.

8. The phases of bacterial growth can be characterized by which one of the following sequences?

   (A) Lag, exponential, stationary, death

   (B) Lag, exponential, death, stationary

   (C) Exponential, lag, death, stationary

   (D) Stationary, exponential, lag, death

   (E) Exponential, stationary, death, lag

9. Which is the most sensitive method of measuring growth of bacterial populations?

   (A) Viable count

   (B) Direct microscope counts

   (C) Dry weight measurement

   (D) Turbidity

   (E) Most probable number

10. Which one of the following growth conditions accounts for the doubling of cellular components when a cell doubles?

    (A) Synchronous growth

    (B) Cryptic growth

    (C) Exponential growth

    (D) Maximum growth

    (E) Balanced growth

ANSWER KEY

1.	(E)	6.	(C)
2.	(D)	7.	(C)
3.	(C)	8.	(A)
4.	(B)	9.	(A)
5.	(B)	10.	(E)

CHAPTER 7

Control of Microbial Growth—Disinfection and Antisepsis

7.1 General Terms

There are various ways in which microbial growth can be retarded or inhibited. Some methods are as follows:

Sterilization—the process of killing (or removing) all microorganisms on an object or in a material (e.g., liquid media).

Disinfection—the process of reducing the numbers of or inhibiting the growth of microorganisms, especially pathogens, to the point where they no longer pose a threat of disease.

Disinfectant—a chemical agent used to destroy microorganisms on inanimate objects such as dishes, tables, and floors. Disinfectants are not safe for living tissues.

Antiseptic—a chemical agent that can be administered safely to external body surfaces or mucous membranes to decrease microbial numbers. Antiseptics cannot be taken internally.

-static agents—those that inhibit growth of microorganisms but do not kill them. A **bacteriostatic** agent is one that inhibits bacterial growth.

-cidal agents—those that kill microorganisms. A **bactericide** is a chemical agent that kills bacteria. A **viricide** is an agent that inactivates viruses. A **fungicide** is an agent that kills fungi. A **sporicide** is an agent that kills spores (bacterial or fungal).

Germicides—are broad-spectrum cidals, including both antiseptics and disinfectants.

Equivalent treatments—different methods or agents that produce the same results with regard to degree of antimicrobial capability.

Selective toxicity—term used to describe the activity of antimicrobial agents that are more harmful to the microbes than they are to the host—a desirable trait.

Problem Solving Example:

 Why does milk "spoil" when kept in a refrigerator?

Even though bacteria may be provided with the proper nutrients for cultivation, it is necessary to determine the physical environment in which they will grow best. Bacteria exhibit diverse reactions to the temperature of their environment. The process of growth is dependent on chemical reactions, and the rates of these reactions are influenced by temperature. Temperature therefore affects the growth rate of bacteria.

Most bacteria grow optimally within a temperature range of 25 to 40°C. (The normal temperature of the human body is 37°C.) These bacteria are termed mesophiles. There are bacteria that grow best at temperatures between 45 and 60°C. These bacteria are termed thermophiles. Some thermophiles will not grow at temperatures in the mesophilic range. At the opposite end of this thermal spectrum are the psychrophiles, bacteria that are able to grow at 0°C or lower. Most psychrophiles grow optimally at higher temperatures of about 15 to 20°C. The psychrophiles are responsible for the spoilage of milk in the cool temperatures of refrigerators (about 5°C). After a week or so, pasteurized milk will begin to "spoil." The accumulation of metabolic products of psychrophilic bacteria will impart an abnormal flavor or odor to the milk. The milk might become viscous, which is a condition referred to as "ropy" fermentation. The viscosity is caused by the accumulation of a gumlike material that normally forms a capsule around each bacterium. Sweet curdling may also occur, caused by the coagulation of casein, a milk protein.

7.2 Factors Influencing Disinfectant Activity

Disinfectant activity is affected by the number of microorganisms, the species and types of microbes (some are more resistant than others, e.g., gram-positive bacteria are generally more sensitive to antibiotics than are gram-negative ones), physiology of organisms (growing organisms are more susceptible than dormant ones), environment (pH, presence or absence of organic matter), and temperature (increased temperatures generally enhance disinfectant activity).

Most antimicrobial agents exert their effect by damaging either the plasma membrane or proteins or nucleic acids.

Problem Solving Example:

 Explain why *Bacillus* and *Pseudomonas* are resistant to disinfectants.

Pseudomonas have structural openings in the cell wall called porins. Porins form channels in the cell wall that allow the passage of small molecules through the cell wall. For this reason, *Pseudomonas* are usually resistant to disinfectants, antiseptics, and some antibiotics. *Pseudomonas* are therefore a common problem in hospitals. New types of plastic and metal medical instruments are being developed that will prevent contamination by this bacterium.

The genus *Bacillus* is an endospore forming gram-positive rod. Endospores are dormant structures with thick walls and additional layers formed internally within a bacterium. Endospores are released into the environment. Endospores of bacteria are highly resistant to chemical disinfectants, as well as extreme temperatures, dehydration, and radiation exposure.

7.3 Physical Methods

Heat is an economical and simple way to destroy microbes. All heat methods work by denaturing proteins.

Thermal death point (TDP)—the lowest temperature at which all bacteria in a liquid culture are killed within 10 minutes.

Thermal death time (TDT)—the time required to kill all bacteria in a liquid culture at a given temperature.

Decimal reduction time (DRT)—the time required to kill 90% of the bacteria in a liquid culture at a given temperature.

Moist heat methods include boiling, pasteurization, and autoclaving.

Boiling—very inexpensive and readily available; usually 100°C for 15 minutes—many vegetative cells and viruses are killed/inactivated within 10 minutes at 100°C.

Temperatures between 0° and 7°C may inhibit the reproduction of certain organisms or the production of toxins. However, these temperatures are rarely bactericidal. Freezing also may not be an effective method of disinfection. In fact, quick freezing is often used to store microorganisms for long periods of time. Slow freezing, however, causes severe damage to cellular constituents and may be bactericidal.

Pasteurization—primarily used to decrease the number of pathogenic organisms in food without adversely affecting the flavor; usually 72°C for 15 minutes or 63°C for 30 minutes.

Autoclaving—steam under pressure—the most effective moist heat method; usually 121.5°C at 15 psi for 15 minutes.

Dry heat methods of sterilization include direct flaming or incineration and hot air (160°C–170°C).

Desiccation—drying or freeze-drying can be used to inhibit growth (via inhibition of enzymes); organisms remain viable.

Osmotic pressure—extremely hypertonic conditions can cause plasmolysis (i.e., contraction of all the cell membrane away from the cell wall).

Filtration is a mechanical means of removing microorganisms. The liquid or gas is passed through a filter with pores small enough to prevent passage of microbes. This method can be used for substances that are sensitive to heat.

The effect of **radiation** is dependent on wavelength and on intensity and duration of exposure.

Ionizing radiation (alpha, beta, gamma, and x-rays, cathode rays, high-energy protons and neutrons) exhibits a high degree of penetrance. It creates free radicals in the medium, leading to the denaturation of proteins and nucleic acids. It can result in mutations. Viruses and spores are somewhat resistant. Gram-negative bacteria are more sensitive to ionizing radiation than are gram-positive organisms.

Ultraviolet radiation is a form of nonionizing radiation. There is a low degree of penetration. It results in thymine dimers (cross-linkages) in DNA that interfere with replication.

Microwaves do not kill organisms directly, but they may be killed indirectly from heat generated in microwaved materials.

Visible light can cause oxidation of some light-sensitive materials.

Problem Solving Example:

 There is a disease of tobacco plants called tobacco mosaic disease. Explain how one can demonstrate the causative agent.

 Tobacco mosaic disease causes the leaves of the tobacco plant to become wrinkled and mottled. By grinding these leaves, one can extract the juice from the infected plant. If this juice is rubbed onto the leaves of a healthy plant, it becomes infected. The agent can therefore be transmitted in the juice.

If the juice is boiled before being rubbed on a healthy plant, no disease develops. The agent might therefore be a bacterium, since most bacteria cannot survive temperatures above 70°C. To isolate the bacteria, we could use filters with known pore diameters. Since most bacteria are larger than 0.5 μ (a micron, μ, is equal to 10^{-6} meter), we could use a filter with a pore size slightly less than this. The liquid is passed through this very fine filter in order to remove the bacteria, and the filtrate is checked for the absence of bacteria. When the filtrate is rubbed on the leaves of a healthy plant, it still causes infection. The causative agent could therefore be a toxin produced by some bacteria or a bacterium smaller than any known. It can be demonstrated that the agent is not a toxin by showing that the agent can reproduce. If the filtrate is used to infect a healthy plant, and this plant is then ground up to obtain a new filtrate, this new filtrate could be used to infect another plant. If this procedure is repeated, and if the extent of mottling becomes decreased with subsequent infections, the agent is most probably a toxin. The attenuation of the disease can be attributed to the dilution of the toxin. However, in tobacco mosaic disease the extent of the mottling does not decrease with repetitions of filtration. One can assume that the causative agent does not become diluted. Another process must be occurring, since the agent is reproducing. Further experiments can show that the causative agent reproduces itself only inside the living plant; it cannot grow on artificial media. It is therefore not a bacterium, since bacteria do not require living host cells to reproduce.

The microbial agent of this disease is termed a virus, the tobacco mosaic virus (TMV). It can be isolated and crystallized and observed using the electron microscope. It is a rod-shaped virus, much smaller than any known group of bacteria: 0.28 μ in length, 0.015 μ in diameter.

7.4 Chemical Disinfection and Sterilization

Chemical methods are often referred to as **cold sterilization**. However, very few actually achieve sterilization.

The chemical structure of chlorhexidine is similar to hexachlorophene. **Chlorhexidine** is often combined with detergents or alcohol as a disinfectant of skin. It is an effective disinfectant of most vegetative bacteria and enveloped viruses. It is used in surgical hand scrubs.

Quaternary ammonium compounds or **quats** are cationic detergents. They are widely used. They are bactericidal against gram-positive bacteria, but less effective against gram-negative bacteria. Quats are also fungicidal, amoebicidal, and are effective against enveloped viruses. The mechanism of action of these detergents is unknown. However, the permeability of the membrane is probably affected, and they may also denature proteins.

7.5 Evaluating a Disinfectant

An ideal disinfectant quickly kills microorganisms without causing damage to the contaminated material. **Potency** is affected by concentration of the agent, length of exposure, temperature, and pH. Evaluation is difficult, but methods do exist, including the following:

Phenol coefficient test—compares the activity of a given agent relative to the killing power of phenol for the same amount of time under identical conditions.

Use-dilution test—rates agents by strength at various dilutions. A chemical that can be greatly diluted and still be effective gets a higher rating. A use-dilution is a dilution that kills all microorganisms at the 95% level of confidence.

Direct-spray method—used to test chemicals that are not water-soluble.

Tissue-toxicity test—tests antiseptics through exposure of tissue culture systems to dilutions of the agent.

Table 7.1 Chemical Disinfectants

Chemical Agent	Action	Examples
Phenolics	Very toxic, disrupt cell membranes and denature proteins	Phenol, cresol, hexachlorophene
Alcohols	Disrupt membranes and denature proteins	Ethanol, methanol, isopropanol
Aldehydes (alkylating agents)	Very effective, denature proteins	Formaldehyde, glutaraldehyde
Oxidizing agents	Very toxic to humans, oxidize molecules within cells, generate oxygen gas	Ozone, peroxide
Halogens	Negatively affected by presence of organic matter, oxidize cell components, disrupt membranes	Iodine, chlorine
Heavy metals	Inactivated by organic compounds, combine with sulfhydryl groups, denature proteins	Silver, mercury (very toxic), copper, zinc, selenium, arsenic
Surface-acting agents	Vary in degree, can simply reduce surface tension allowing organisms to be washed away, or may disrupt membranes and denature proteins	Soaps, detergents (including quaternary ammonium compounds), surfactants
Organic acids	Inhibit fungal metabolism (used as food preservatives)	Benzoic acid, propionic acid, sorbic acid
Gases	Denature proteins	Ethylene oxide (very toxic), vapors from formaldehyde, methyl bromide
Antiseptic dyes	Block cell wall synthesis, interfere with DNA replication	Acriflavine, crystal violet

Problem Solving Example:

Q A student uses the use-dilution test for two different disinfectants, A and B. The values for Disinfectant A is 1:50 and Disinfectant B is 1:10,000. Disinfectants are tested under identical conditions. Which is the more effective disinfectant?

A Disinfectant B is more effective because it has a higher use-dilution test value. This value indicates the amount of dilution needed to the same degree of disinfection. Under the conditions of this experiment, less of Disinfectant B is needed to achieve the same amount of disinfection.

7.6 Microbial Death

Bacteria that are treated with physical methods of microbial control or antimicrobial chemicals tend to die at a constant rate. When plotted on a semilogarithmic graph of log numbers of survivors versus time, the result is a straight line. Therefore, bacterial death is constant.

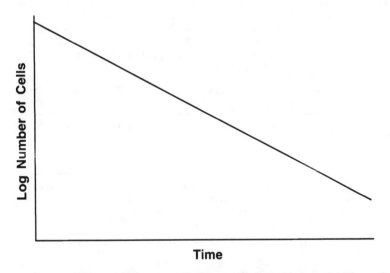

Fig. 7.1 An Exponential Plot of the Surviving Cells Versus the Time of Exposure to a Lethal Agent

Problem Solving Example:

Q Illustrate the result of adding (a) an antibacterial compound to a bacterial culture in log phase and (b) a bacteriostatic compound to a bacterial culture in log phase.

A A bacterial culture in log phase is one that is growing at an exponential rate. An antibacterial compound is an agent that kills microorganisms or inhibits their growth. This is why on the graph the number of cells decreases with time once the antimicrobial agent is added to the culture. A bacteriostatic compound inhibits the growth and reproduction of microorganisms, but does not kill them. This is why on the graph the cell number does not decrease after the bacteriostatic compound is added, but remains constant over time.

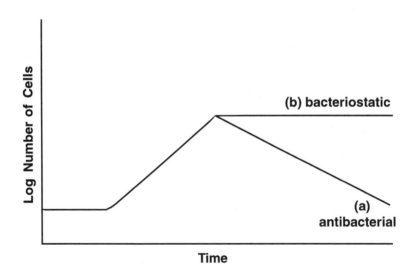

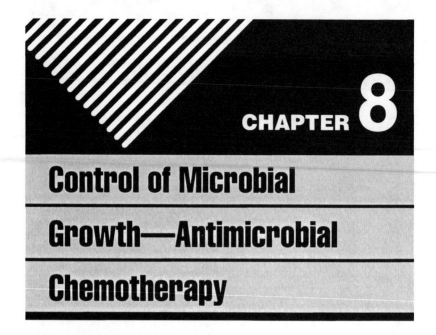

CHAPTER 8

Control of Microbial Growth—Antimicrobial Chemotherapy

8.1 General Terms

The ideal antimicrobial agent should be nontoxic to the host (selective toxicity), nonallergenic, soluble in body fluids, able to be maintained at therapeutic levels, have a low probability of eliciting resistance, long shelf life, and low cost.

Antibiotic—originally used to denote a chemical substance produced by one microorganism that kills or inhibits the growth of other microbes, the term now applies to both naturally produced substances and those synthesized in the laboratory. Most are produced by either fungi (e.g., penicillins, cephalosporins), *Bacillus* species (e.g., polymyxin, bacitracin), or *Streptomyces* species (streptomycin, tetracycline, erythromycin, kanamycin, neomycin, nystatin). Broad-spectrum antibiotics are those that act on both gram-positive and gram-negative bacteria.

Chemotherapeutic agent (drug)—any chemical (natural or synthetic) that is used in medicine. Ideally, it should attack microorganisms selectively and not harm human cells.

Natural drug—one made by microorganisms. **Synthetic drug**—one that is made in the laboratory. **Semisynthetic drug**—one synthesized partly in the laboratory and partly by microorganisms.

Synergistic effect—antibiotic effectiveness is often enhanced when given in combination with another drug. On the other hand, antagonistic effect is sometimes observed wherein certain combinations hinder antibiotic effectiveness.

Minimal inhibitory concentration (MIC)—the lowest drug concentration that will prevent growth of a standardized microbial suspension.

Minimal antibacterial (active) concentration (MAC)—the amount of drug (generally extremely small) that causes changes in cell morphology and inhibits growth.

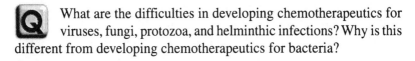

Problem Solving Example:

Q What are the difficulties in developing chemotherapeutics for viruses, fungi, protozoa, and helminthic infections? Why is this different from developing chemotherapeutics for bacteria?

A Bacteria are prokaryotic cells. It is relatively easy to find a drug that will affect the prokaryotic cell and not the eukaryotic cell of the host because the two cell types differ in cellular structure. A primary target for therapeutics is the bacterial cell wall.

Viruses are obligate cellular parasites and therefore require a eukaryotic host cell to complete their life cycle. Antiviral drugs must recognize eukaryotic cells that have been infected by a virus and then destroy these infected cells.

Fungi, protozoa, and helminths are eukaryotes. These eukaryotes resemble the host's cell structurally. Antifungals, antiprotozoans, and antihelminthics must distinguish between the pathogen and the host eukaryotic cells in order to be effective.

8.2 Types of Agents

8.2.1 Antimycotics

Antimycotics (antifungals)—drugs that affect fungal growth, generally through inhibition of cell division, disruption of nucleic acid and protein synthesis, or changes in the cell membrane. Antifungal drugs include amphotericin B, clotrimazole, miconazole, and nystatin.

Polyene antibiotics are produced by a species of *Streptomyces* and are effective antifungal agents against opportunistic mycoses. These drugs combine with the sterols already present in the fungal membrane to change the permeability of the membrane. The result is cell death. These antibiotics are not effective against organisms that do not have sterols in their membranes and are therefore not bactericidal against them (mycoplasmas are the exception). **Nystatin** is used against *Candida* infections.

Clotrimazole and **miconazole** are **imidazole antifungals**, which may be used against cutaneous mycoses such as athlete's foot. **Ketoconazole** is an imidazole antifungal, which is used against systemic mycoses. These imidazole antifungals interfere with sterol synthesis and inhibit the formation of a fully functional fungal membrane.

Triazole antifungals include **fluconazole** and **itraconazole**. Both are used for systemic mycoses.

Griseofulvin is produced by a *Penicillium* and interferes with mitosis. It is effective when taken orally against superficial dermatophytic mycoses of the hair. It acts by disrupting cell division.

8.2.2 Antiprotozoals

Antiprotozoals—drugs that affect the growth of protozoans; often toxic and not always easy to obtain. Examples include **quinacrine hydrochloride** and **metronidazole**.

8.2.3 Antivirals

Antiviral agents must be shown to be nontoxic to humans before they can be used. The **interferons** (IFNs) are a group of antiviral and antitumor proteins. They are released from virus-infected cells, causing neighboring cells to make antiviral proteins. Interferons inhibit viral replication. Recombinant DNA technology is used to produce high levels of interferons. The three types are alpha, beta, and gamma IFNs. Other antiviral agents include **acyclovir**, **amantadine**, and **5-iodo-2′-deoxyuridine**.

8.2.4 Antihelminthics

Niclosamide is an **antihelminthic drug** used against the beef tapeworm *(Taenia saginata)*. **Praziquantel** is used against tapeworms as well as some flukes, including *Schistosomas*. **Mebendazole** is used against *Ascaris lumbricoids, Enterobius vermicularis*, and *Trichuris trichiura*. **Piperazine** is used to paralyze pinworms and ascariasis.

8.3 Mechanisms of Action of Antimicrobial Drugs

Cidal drugs cause irreversible damage or death and are independent of host activity. **Static drugs** inhibit growth or reproduction and are dependent on the immune system of the host for elimination of pathogens from the body.

Antimicrobial agents can inhibit growth or kill microbes through inhibition of nucleic acid synthesis, protein synthesis, cell wall formation, or metabolic products or by damaging the plasma membrane, causing cell lysis.

8.3.1 Cell Wall Synthesis

Peptidoglycan is unique to bacterial cell walls. Antibiotics that prevent the synthesis of peptidoglycans weaken the cell wall of the bacteria and cause cell lysis. Penicillin prevents the synthesis of peptidoglycans and is effective against gram-positive bacteria. The lipopolysaccharides of the outer membrane of gram-negative bacteria allow these bacteria to be resistant to these types of antibiotics. Bacitracin, vancomycin, cephalosporins, and penicillin are antibiotics that interfere with the synthesis of peptidoglycans. These antibiotics do not affect the eukaryotic host cells because eukaryotic cells do not contain peptidoglycans.

8.3.2 Protein Synthesis

Chloramphenicol, erythromycin, streptomycin, and tetracyclines are antibiotics that interfere with **protein synthesis**. These antibiotics interfere with the bacterial ribosomes involved in the process of synthesizing a protein strand. These antibiotics do not affect the eukaryotic cells because eukaryotes use different ribosomes in protein synthesis.

8.3.3 Plasma Membranes

Disruption of the cell membrane causes a change in the permeability of the cell and can cause cell lysis. Polymyxin B attaches to the phospholipids of the membranes. Nystatin, amphotericin B, miconazole, and ketoconazole are effective antifungals that selectively cause the increased permeability of the fungal membrane because these attach to the sterols in the fungal membrane. Animal cells also contain sterols; therefore, some of these drugs may be toxic to the host.

8.3.4 Replication

Idoxuridine, rifamycin, and quinolones are antibiotics that interfere with **DNA replication** and **transcription**. However, these may also interfere with host replication and transcription. Selective toxicity is vital in the effectiveness of the antibiotic.

8.3.5 Synthesis of Essential Metabolites

Antibiotics that compete with microbial enzymes may be effective chemotherapies. Sulfonamides and dapsone inhibit the synthesis of folic acid in many microorganisms. The inhibition of folic acid synthesis causes the microorganism to stop growing. Humans do not produce folic acid by this metabolic pathway. Human host cells are therefore not affected by these drugs. Trimethoprim blocks tetrahydrofolate synthesis, and isoniazid is thought to inhibit the synthesis of mycolic acid.

Problem Solving Example:

 Characterize the following commonly used antimicrobial drugs:

    (a) Isoniazid

    (b) Cephalosporin

    (c) Imidazole

    (d) Amantadine

    (e) Metronidazole

    (f) Praziquantel

(a) Isoniazid is a synthetic antibacterial drug. It is very effective against *Mycobacterium tuberculosis*. Its mechanism of action is inhibition of mycolic acid synthesis. Mycolic acid is a component of the cell wall found exclusively in mycobacteria. Therefore, isoniazid is only effective on mycobacteria.

(b) Cephalosporin is a natural antibacterial drug. Its mechanism of action is inhibition of cell wall synthesis. It is more effective in gram-negative bacteria than the penicillinases and affects a different group of beta-lactamases.

(c) Imidazoles are antifungal drugs and include clotrimazole, miconazole, and ketoconazole in the treatment of cutaneous fungal infections such as athlete's foot. The mechanism of action is interference in sterol synthesis and disruption of fungal membrane permeability. Fluconazole is used to treat systemic fungal infections.

(d) Amantadine is an antiviral drug. The mechanism of action is inhibition of the penetration and uncoating of a virus. It is used in nursing homes to prevent the spread of viral infections.

(e) Metronidazole is an antiprotozoan drug and also is effective against obligate anaerobic bacteria. It is used to treat *Trichomonas vaginalis*, giardiasis, and amoebic dysentery. The mechanism of action is to damage DNA.

(f) Praziquantel is an antihelminthic drug used for treatment of tapeworms and fluke-caused diseases such as schistosomiasis. Praziquantel acts at the membranes of affected helminths to cause muscular contraction resulting in detachment from host tissues. It also causes tegumental damage, which activates host defense mechanisms, thus destroying the worms.

8.4 Evaluating an Antimicrobial Drug

There are a number of methods for testing the sensitivity of microbes to antibiotics; these include **serum killing power, Kirby-Bauer disk diffusion**, and automated techniques.

Serum killing power—tests activity of the antibiotic in the patient's serum. The patient's serum is drawn during the time period when the patient is receiving the antibiotic. A bacterial suspension is added to the serum to see if the microbes are affected.

Disk diffusion method (Kirby-Bauer)—infecting microorganisms are cultured on agar plates and subjected to filter disks that have been impregnated with various antibiotics. Cleared **zones of inhibition** indicate antibiotic sensitivity.

Problem Solving Example:

An antibiotic drug (200 µg/ml) was used in a broth dilution test, and the following results were obtained:

Antibiotic Dilution	Growth	Growth with Antibiotic
not diluted	–	–
1:10	–	–
1:100	–	+
1:1,000	+	+

What is the minimal inhibitory concentration (MIC) and the minimal bactericidal concentration (MBC), also known as the minimal lethal concentration (MLC), of this antibiotic?

 The MIC is determined by the smallest concentration of antibiotic that prevents visible growth. This number is 2 μg/ml (1:100 dilution of 200 μg/ml).

The MBC is 20 μg/ml (1:10 dilution of 200 μg/ml). The cells that do not show growth in the MIC test are cultured in media free of the antimicrobial. If there is growth, then the antimicrobial is not bactericidal. The MBC can then be calculated.

8.5 Side Effects

Drugs vary in effectiveness and in the number and degree of severity of **side effects**. Side effects of antimicrobial agents include toxicity, allergic response, and disruption of normal flora.

Many drugs attack not only the infectious organisms, but the normal flora as well. **Superinfections** with new pathogens can occur when the defensive capacity of the host's normal flora is compromised or destroyed.

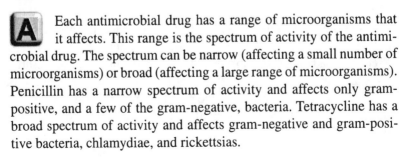

Problem Solving Example:

 Distinguish between the terms: spectrum of activity, broad and narrow spectrum, and superinfection.

 Each antimicrobial drug has a range of microorganisms that it affects. This range is the spectrum of activity of the antimicrobial drug. The spectrum can be narrow (affecting a small number of microorganisms) or broad (affecting a large range of microorganisms). Penicillin has a narrow spectrum of activity and affects only gram-positive, and a few of the gram-negative, bacteria. Tetracycline has a broad spectrum of activity and affects gram-negative and gram-positive bacteria, chlamydiae, and rickettsias.

Broad spectrum antimicrobials affect the normal microorganisms present in the host as well as the pathogenic organisms. However, in some cases a competitor microorganism may not be affected by the antimicrobial drug and will result in overgrowth or superinfection. For example, *Candida albicans* is not affected by bacterial antibiotics. Treatment with a bacterial antibiotic will result in a superinfection of *Candida albicans*.

8.6 Drug Resistance

Resistance to an antibiotic means that a microorganism that was formerly susceptible to the action of that antibiotic is no longer affected by it. Antibiotic resistance can sometimes be transferred among bacteria on extrachromosomal DNA molecules known as plasmids. Resistance may be due to changes in the sensitivity of affected enzymes, changes in the selective permeability of cell walls and membranes, increased production of a competitive substrate, or enzymatic alteration of the drug itself.

Unnecessary exposure to antibiotics has brought about a significant increase in **antibiotic-resistant microbes**. This calls into question the routine uses of antibiotics such as feed additives for livestock and food additives used to prolong freshness in agricultural products. These practices create conditions that are conducive to the development of resistant strains that may be transferred to a human host when the product is ingested.

Nosocomial infections are infections acquired during a hospital stay. They are often extremely resistant to antibiotics and very difficult to treat.

A group of gram-negative bacteria are resistant to disinfectants and antiseptics; they are the **Pseudomonads** *(genus Pseudomonas)*. Pseudomonads may even thrive and grow in the presence of disinfectants and antiseptics. The resistance of these bacteria is thought to be due to the presence of **porins** in the membrane. Porins are proteins that form channels, which permit small molecules to flow into and out of the cell.

Mycobacterium tuberculosis, the causative agent in **tuberculosis**, is also resistant to disinfectants, as well as antibiotics.

Endospores and cysts of bacteria and protozoa, respectively, are also resistant to disinfectants.

Problem Solving Example:

 How do bacteria develop drug resistance?

Most antibiotic-resistant bacteria result from genetic changes and subsequent selection. The genetic changes may be due to chromosomal mutations or to the introduction of extrachromosomal elements.

Spontaneous mutations in a bacterial chromosome can cause antibiotic resistance in several forms. The mutation may make the cell impermeable to the drug by changing the shape of the receptor molecule. The mutation may create an enzyme that inactivates the drug once it enters the cell. The mutation may make the drug's intercellular targets resistant to the drug. Streptomycin, which inhibits the binding of formylmethionyl tRNA to the ribosomes, may be blocked if the ribosome was changed so that the interaction was prevented.

Antibiotic resistance may also arise extrachromosomally. Conjugal plasmids, such as R plasmids, contain genes that mediate their genetic transmission. R plasmids carry genes conferring antibiotic resistance. Thus, R$^+$ cells can pass the genes for resistance to R$^-$ cells by conjugation.

Once a bacterial cell strain has become resistant to an antibiotic, the presence of that antibiotic in the environment favors the cells that contain the resistance element. Cells without the resistance will be killed by the antibiotic; those that have the resistance will flourish.

8.7   Some Common Antibacterial Drugs

Penicillins (including ampicillins and cephalosporins)—narrow-spectrum bactericides; interfere with cell wall formation; effective only against vegetative cells of gram-positive sensitive species. Susceptible to beta-lactamase activity.

Some organisms, such as *Haemophilus influenzae*, *Neisseria gonorrhoeae*, and *Staphylococcus aureus*, produce an enzyme known as beta-lactamase. The presence of this enzyme makes the organisms resistant to beta-lactam antibiotics—namely, penicillins and cephalosporins.

Monobactams—effective against aerobic, gram-negative organisms (e.g., *Enterobacter*, *Haemophilus*, *Neisseria*); stable to beta-lactamase.

Chloramphenicol and **tetracyclines**—broad-spectrum bacteriostatics that act on both gram-positive and gram-negative bacteria; inhibit protein synthesis.

Aminoglycosides (including kanamycin and streptomycin)—bactericidal; interfere with protein synthesis.

Polymyxin B and **colistin**—disrupt the plasma membrane, allowing cytoplasmic constituents to leak out.

Sulfa drugs—bacteriostatic; active upon vegetative cells; competitively inhibit enzymes.

Problem Solving Example:

Q Explain why penicillin is effective only against actively growing bacteria. Describe the mode of action of some other antimicrobial agents.

A An antimicrobial agent is one that interferes with the growth and activity of microorganisms. Knowledge of the mode of action of a particular agent makes it possible to determine the conditions under which it will act most effectively. Among the known sites of action of antimicrobial agents are the cell wall, cell membrane, protein structure and synthesis, and enzyme activity.

The cell walls of some gram-positive bacteria are attacked by the enzyme lysozyme, which is normally found in tears, mucous secretions, and leukocytes (white blood cells). Lysozyme breaks down the peptide linkages in the cell wall complex. Some bacteria secrete enzymes that degrade cell walls of other bacteria or prevent cell wall formation. Without a cell wall to provide support for the bacterium, it will soon lyse and die. The antimicrobial effect of penicillin is attributed to its inhibition of cell wall synthesis. Penicillin prevents the incorporation of the amino sugar N-acetylmuramic acid into the mucopeptide structure that comprises the cell wall. This is why penicillin works only on actively growing bacteria. If cell wall formation is complete, penicillin has no effect.

The cell membrane helps contain the cellular constituents and provides for selective transport of nutrients into the cell. Damage to this membrane thus inhibits growth or causes death. The bactericidal (kills bacteria) action of phenolic compounds, such as hexachlorophene, is attributed to their effect on cell permeability. This results in leakage of cellular constituents and eventual death.

Proteins are essential to the cell for both structure and enzymatic activity. Protein denaturation (alternation of their natural configuration) causes irreparable damage to the cell. High temperatures, acidity, and alcohol denature proteins. Streptomycin combines with the ribosomes of sensitive bacteria and disturbs protein synthesis.

Many agents inhibit enzymes involved in the energy-supplying reactions of the cell. For example, cyanide inhibits cytochrome oxidase in the electron transport chain, fluoride inhibits glycolysis, and dinitrophenol uncouples oxidative phosphorylation. All of these inhibit ATP synthesis.

A familiar antibiotic, sulfanilamide, works by blocking the synthesis of folic acid, a necessary substrate for certain reactions in the cell. A precursor of folic acid is paraaminobenzoic acid (PABA) whose structure is very similar to that of sulfanilamide.

Sulfanilamide, by effectively competing with its precursor for the binding site on an enzyme involved in the pathway, inhibits folic acid synthesis. Any compound, such as sulfanilamide, that interrupts synthetic processes by substituting itself for a natural metabolite is called an antimetabolite or metabolic analogue.

Quiz: Control of Microbial Growth

1. Carla, curious as to the cause of disease in her mother's tobacco plants, designed the following experiment. She crushed the infected leaves with sterile water. Then, the solution was filtered once through a funnel filter and once through a 0.4 micron filter. Both the filtrate and residue of each filter were kept and tested by rubbing each, individually, onto an uninfected plant. The table below shows the results:

	Infection Ensues	No Infection
Initial filtrate	X	
Initial residue	X	
Second filtrate	X	
Second residue		X

Which one of the following is the most likely pathogen?

    (A) Virus

    (B) Bacteriophage

    (C) Bacterium

    (D) Fungus

    (E) None of the above.

2. Which substance produced by a microorganism is used to kill or inhibit the growth of other microorganisms?

    (A) Antibiotic

    (B) Antibody

(C) Antigen

(D) Complement

(E) Interferon

3. Numerous antibiotics work by interfering with protein synthesis. An example is cycloheximide, which

    (A) inhibits the peptidyl transferase activity of 60S ribosomal subunits in eukaryotes.

    (B) binds to the 30S subunit and inhibits binding of aminoacyl-tRNAs in prokaryotes.

    (C) causes premature chain termination by acting as an analogue of aminoacyl-tRNA in both prokaryotes and eukaryotes.

    (D) inhibits translocations by binding to the 50S ribosomal subunit in prokaryotes.

    (E) inhibits initiation and causes misreading of mRNA in prokaryotes.

4. The material responsible for conferring resistance to antibiotics is the

    (A) F⁻ factor.

    (B) Hfr factor.

    (C) Col plasmid.

    (D) R plasmid.

    (E) F' factor.

5. How are penicillin and lysozyme similar?

    (A) Both consist of amino acids.

    (B) Both are antibiotics.

    (C) They are the same.

(D) Both affect cell walls.

(E) Both affect the beta-lactamase.

6. Culture media are sterilized by

   (A) filtration.

   (B) autoclaving.

   (C) antibiotics.

   (D) chemical treatment.

   (E) desiccation.

7. In order for an antibiotic to be effective, it must

   (A) have a broad spectrum.

   (B) have a narrow spectrum.

   (C) have selective toxicity.

   (D) have a synergistic effect.

   (E) be semisynthetic.

8. Phenolic compounds

   (A) are alkylating agents.

   (B) combine with sulfhydryl groups.

   (C) are oxidizing agents.

   (D) are protein precipitants.

   (E) cause protein denaturation.

9. Which one of the following is/are resistant to disinfectants?

   (A) Pseudomonads

   (B) Endospores

   (C) *Mycobacterium tuberculosis*

(D) Cysts

(E) All of the above.

10. Pasteurization is used in the preparation of which of the following commercial food preparations?

   (A) Wine

   (B) Beer

   (C) Milk

   (D) Orange juice

   (E) All of the above.

ANSWER KEY

1.	(A)	6.	(B)
2.	(A)	7.	(C)
3.	(A)	8.	(E)
4.	(D)	9.	(E)
5.	(D)	10.	(E)

CHAPTER 9

Microbial Genetics

9.1 Genetics

Genetics is the study of genes, what they are, and how they work, including how they store, express, and replicate information, and how that information is transmitted to subsequent generations (**heredity**). A **gene** is a segment of DNA, composed of a sequence of nucleotides that specify the structure of a functional product, usually a protein.

DNA—deoxyribonucleic acid, the genetic substance of a cell. DNA exists as a double-stranded, helical molecule (see figure 9.1). The two strands are composed of a series of nitrogenous bases (A = adenine, T = thymine, G = guanine, and C = cytosine) that are connected by sugar-phosphate molecules. The nitrogenous bases opposing each other along the two strands are linked by hydrogen bonds (adenine pairs with thymine; guanine with cytosine). Each base pair is composed of one purine (A or G) and one pyrimidine (T or C).

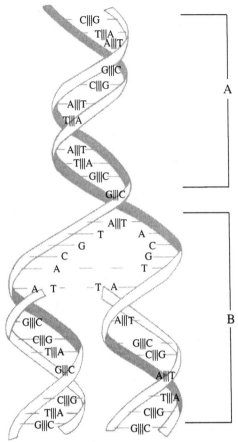

Figure 9.1 DNA. The double helix (A) is found in an undividing eukaryotic cell. The lower section (B) has been uncoiled, and each template strand is being replicated.

Problem Solving Example:

The central dogma of biochemical genetics is the basic relationship between DNA, RNA, and protein. DNA serves as a template for both its own replication and synthesis of RNA, and RNA serves as a template for protein synthesis. How do viruses provide an exception to this flow scheme for genetic information?

The central dogma of biochemical genetics can be summarized in the following diagram:

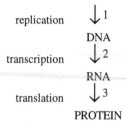

replication   ↓ 1

DNA

transcription   ↓ 2

RNA

translation   ↓ 3

PROTEIN

Arrow 1 signifies that DNA is the template for its own replication. Arrow 2 signifies that all cellular RNA molecules are made on DNA templates. All amino acid sequences in proteins are determined by RNA templates (arrow 3). However, in the viral infection of a cell, RNA sometimes acts as a template for DNA, providing an exception to the unidirectionality of the scheme involving replication, transcription, and translation. The agents involved are certain RNA viruses. For example, an RNA tumor virus undergoes a life cycle in which it becomes a prophage integrated into the DNA of a host chromosome. But how can a single-stranded RNA molecule become incorporated into the double-helical structure of the host DNA? It actually does not.

An RNA tumor virus first absorbs to the surface of a susceptible host cell, then penetrates the cell by an endocytotic (engulfment) process, so that the whole virion (viral particle) is within the host cell. There, the particle loses its protein coat (probably by the action of cellular proteolytic enzymes). In the cytoplasm, the RNA molecule becomes transcribed into a complementary DNA strand. The enzyme mediating this reaction is called reverse transcriptase and is only found in viruses classified as retroviruses. (Thus, cells do not have the capacity to transcribe DNA from an RNA template.) Cellular DNA polymerase then converts the virally produced single-stranded DNA into a double-stranded molecule. This viral DNA forms a circle and then integrates into the host chromosome, where it is transcribed into the RNA needed for new viral RNA and also for viral-specific protein synthesis. The RNA molecules and protein coats assemble, and these newly made RNA tumor viruses are enveloped by sections of the cell's outer membrane and detach from the cell surface to infect other cells. The release of the new virions does not require the lysis of the host cell and is accomplished by an evagination of the outer membrane. Other cells are infected through the fusion of the envelope and the potential host cell's outer membrane, thereby releasing the virion into the cytoplasm. Unlike the λ phage infection of an *E. coli* cell, the RNA tumor virus does not necessarily interfere with normal cellular processes and cause death.

Thus, the enzyme reverse transcriptase of RNA viruses provides the basis for the single known exception to the standard relationship between DNA, RNA, and proteins by catalyzing the transcription of DNA from RNA. The virus causing AIDS (acquired immune deficiency syndrome) is a retrovirus and thus utilizes the reverse transcriptase enzyme.

9.2 Chromosomes

A **chromosome** is a single DNA molecule, composed of one long double helix and containing all or a part of the genome. Eukaryotic chromosomes have proteins known as **histones** associated with them; these proteins regulate gene activity. Prokaryotic chromosomes do not have these associated proteins and are said to be **naked**.

Prokaryotes have a single, circular chromosome. Asexual reproduction leads to two daughter cells—each receiving a chromosome that is identical to the original parental chromosome.

Plasmids are circular, extrachromosomal DNA molecules that are generally not essential for cell survival.

Eukaryotes have multiple, linear chromosomes. Reproduction may be asexual, with each new cell receiving chromosomes identical to the parent cell, or sexual, in which the new cell receives a subset of genetic information from each parent.

Problem Solving Example:

Compare and contrast the bacterial chromosome and the eukaryotic chromosome.

A   The nucleic acid of both the bacterial (or prokaryotic) and eukaryotic chromosome is exactly the same: DNA. However, the structure of the DNA is different. Bacteria contain a single circular chromosome that consists of a double-stranded DNA molecule. For example, in *Escherichia coli*, the chromosome contains about 4 million base pairs. The chromosome is tightly packed inside the nucleoid region of the prokaryotic cell.

The eukaryotic chromosome is also tightly packed; however, it is located in the eukaryotic nucleus. The eukaryotic condensed chromosome also contains proteins called histones. Eukaryotic chromosomes are linear.

9.3   Replication

For DNA replication to occur, the two strands of the double helix must separate. The point where they are separated is called the **replication fork**. Each separated strand serves as a **template** for **DNA polymerases** in the synthesis of a new strand. Because of the rules of nitrogenous base pairing (i.e., A pairs only with T, and G pairs only with C), the sequence along the old strand dictates precisely the sequence along the new strand. Each new double-stranded DNA molecule contains one old strand and one new strand; thus, the process of DNA replication is said to be **semiconservative**.

Problem Solving Example:

Q   When ^{15}N is added to the environment of bacteria, 50% of the nitrogen in the DNA of the first new generation of bacteria is ^{15}N. If this generation were isolated from ^{15}N and only exposed to ^{14}N, what isotopic makeup of nitrogen would you expect to find in the second generation of bacteria? (Assume that there is no ^{15}N in the bacterial environment.)

    Half of the second generation bacteria have 50% of their DNA's nitrogen in the ^{15}N form, and half have no ^{15}N in their DNA.

The first new generation is reproducing on ^{14}N. The progeny will have one-half of its DNA newly synthesized on ^{14}N, and half of its DNA already present. From this it can be inferred that half of the second new DNA generation would contain only ^{14}N strands, and the other half would contain 50% ^{14}N and 50% ^{15}N.

9.4 Transcription—Synthesis of RNA

Transcription is the synthesis of single-stranded RNA (ribonucleic acid) molecules based on a DNA sequence. The two strands of the DNA must pull apart temporarily, allowing **RNA polymerase** to access the DNA for use as a template for RNA production.

RNA is synthesized from adenine (A), guanine (G), cytosine (C), and uracil (U). Thymine is not present in RNA. The As, Gs, Cs, and Ts along the DNA template produce Us, Cs, Gs, and As, respectively, along the RNA strand being synthesized.

Three kinds of RNA may be produced: (1) **ribosomal RNA (rRNA)**, which combines with proteins to form ribosomes (where new proteins are synthesized), (2) **transfer RNA (tRNA)**, which transports amino acids to the ribosome for assembly into proteins, and (3) **messenger RNA (mRNA)**, which dictates the sequence of amino acid assembly.

Transcription occurs in the cytoplasm of prokaryotes, while it occurs in the nucleus of eukaryotes.

Eukaryotic mRNA contains regions that are not used for protein synthesis (the **introns**), as well as those regions that are (the **exons**). Prior to mRNA transport out of the nucleus, enzymes remove the introns, and the **spliceosome** connects the exons into a functional mRNA that is then exported to the cytoplasm.

Problem Solving Example:

Explain how supercoiling in bacteria facilitates the transcription of genes.

A Bacterial DNA, being circular, has no free end for rotation as the DNA helix unwinds during transcription. Positive supercoils accumulate and hinder further unwinding of the DNA helix and therefore transcription. This problem is solved by DNA gyrase, which introduces negative supercoils into the DNA and causes the circular bacterial DNA to be under tension. Local unwinding of a region of the DNA helix by RNA polymerase introduces positive supercoils, which in effect cancel out negative supercoils formed by DNA gyrase. Relaxation of tension is the energetically favorable event that enhances the transcription of DNA.

9.5 Translation

Translation is the process wherein information in the form of nitrogenous bases along an mRNA is translated into the amino acid sequence of a protein.

The sequence of nucleotides along the mRNA is "read" in groups of three; each group of three is called a **triplet** or **codon**. Codons in mRNA pair with anticodons found in tRNA molecules. The triplet **anticodon** is located at one point on the tRNA molecule while the corresponding amino acid is attached to the tRNA at another point. The mRNA and tRNA are brought together at the ribosome. The ribosome moves along the mRNA strand during the synthesis of the polypeptide.

The mRNA codons make up the **genetic code**, which is essentially identical among all living things. The code comprises 64 codons—61 code for amino acids (these are the **sense codons**), while 3 do not code for amino acids but function as stop signals for the translation process (these are the **nonsense codons**). There are 61 codons coding for 21 amino acids; thus, some amino acids are coded for by more than one codon. Hence, the code is said to be **degenerate**. The **start codon** is **AUG** and codes for the amino acid methionine.

A **polypeptide** is a chain of amino acids.

A **polyribosome** is an mRNA with many ribosomes attached.

Due to the absence of a nuclear membrane, prokaryotic translation can begin even before transcription is completed. This is not so in eukaryotic cells, in which transcription occurs in the nucleus and translation occurs in the cytoplasm.

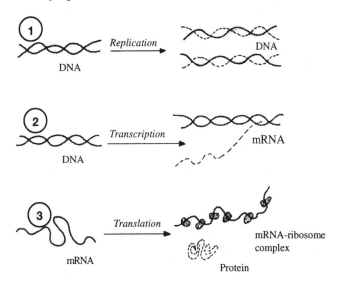

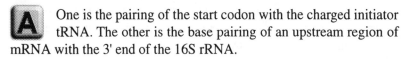

Figure 9.2 Summary: Replication, Transcription, and Translation. (1) This DNA is used in the nucleus of new cells. (2) The mRNA leaves the nucleus and is used in protein synthesis. (3) Amino acids are assembled onto the mRNA-ribosome complex to create a protein.

Problem Solving Example:

Q What are the two kinds of interactions that determine the initiation phase of protein synthesis in bacteria?

A One is the pairing of the start codon with the charged initiator tRNA. The other is the base pairing of an upstream region of mRNA with the 3' end of the 16S rRNA.

The 70S ribosome in prokaryotes consists of a 30S and 50S subunit. The 30S subunit is made up of a 16S rRNA with about 21 proteins, while the 50S subunit consists of a 5S rRNA, a 23S rRNA, and about 34 proteins. The 16S rRNA has near its 3' end a short base sequence complementary to the purine-rich region in the initiation site upstream of the start codon of mRNA. The binding of initiator tRNA to the mRNA-30S ribosomal subunit forms an initiation complex for the subsequent phases of protein synthesis.

9.6 Mutation

Mutations—changes in the sequence of A, T, G, and C along a DNA strand. Mutations are changes in the genotype; they may or may not change the phenotype (the observable characteristics). A **silent mutation** does not alter the phenotype.

Mutations may be **neutral**, **harmful**, or **beneficial** in their effect.

Spontaneous mutations—occur naturally, appear to be due to random errors in base pairing during DNA replication. **Induced mutations** occur as a result of exposure to a **mutagen**—a chemical substance or physical factor that increases the mutation rate; it causes permanent damage to the DNA. Examples include ultraviolet light, nitrous oxide, and benzo(a)pyrene.

Mutation rate—the probability that a mutation will occur in a gene when the cell divides. Different genes have different mutation rates.

Point mutation (base substitution)—change in a single nucleotide; may result in no change in amino acid (**silent mutation**), substitution of a different amino acid (**missense mutation**), or creation of a stop codon (**nonsense mutation**).

Frameshift mutation—a base pair is inserted or deleted; this generally affects the mRNA from that point on, i.e., it can change all triplets for the rest of the sequence. The changes may include silent, missense, and nonsense mutations. Frameshift mutations are more likely to result in a deficient protein product than are point mutations.

9.6.1 Detecting Mutants

Mutants can be identified by **selecting** or testing for an altered phenotype.

Positive selection—mutant cells are selected and nonmutant cells are inhibited.

Negative selection—the **replica plating** technique is used to identify mutants that cannot grow under the selective conditions.

An **auxotroph** is a mutant with a nutritional requirement that was not present in the parent. For example, a histidine auxotroph is a bacterium that cannot grow unless histidine is added to the medium.

9.6.2 Mutagens

The following are examples of **chemical mutagens: base analogs** such as 2-aminopurine, which incorporates in place of thymine; **base-pair mutagens** such as nitrous acid, which converts adenine into hypoxanthine; and frameshift mutagens such as benzo(a)pyrene. Others include alkylating agents, deaminating agents, and acridine derivatives.

Ionizing and ultraviolet radiation are also mutagens. Ionizing radiation can cause base substitutions, disrupt the sugar-phosphate backbone, or create reactive free radicals. Ultraviolet radiation causes thymine dimers, i.e., bonding between adjacent thymines.

Problem Solving Example:

Q Live bacteria, all of which were red in color, were placed under an ultraviolet lamp. After several days, groups of white bacteria began to appear among the red. What conclusions, if any, can be made at this stage?

A At this stage, we can guess that the white bacteria probably arose as the result of a mutation. We know that exposure to ultraviolet radiation usually causes an increase in the mutation rate, i.e., it is said to induce mutations. We also know that the white bacteria arose from among the red, and in the same culture medium. If the original culture was indeed genetically pure (all the individuals having come from the same cell and thus having identical genotypes), there is no reason for any of the bacteria to exhibit different behavior or characteristics while remaining in the same environment. A change in phenotype could only be caused by a change in the genetic structure, in this case, a genetic mutation.

9.6.3 Repair of DNA Damage

Some bacteria have enzymes that can repair DNA. There are two kinds of repair: **light repair**, in which a light-activated enzyme breaks thymine-thymine bonds; and **dark repair**, in which several different enzymes are involved in excising defective DNA and resynthesizing the DNA strand based on the nonmutated strand.

9.6.4 Identifying Carcinogens

The **Ames test** for the identification of carcinogens is based on the assumptions that (1) the presence of a mutagen can cause mutant cells to revert back to their original state, and (2) many mutagens are also carcinogens. The test involves exposure of histidine auxotrophs of *Salmonella* to a suspected mutagen, followed by selection for nonmutant cells. The presence of nonmutant *Salmonella* indicates a positive test for mutagenicity. The Ames test is quick and relatively inexpensive.

Problem Solving Example:

Q How are potential mutagens tested?

A A very powerful way to test potential mutagens is by the Ames test. This procedure involves subjecting a strain of *Salmonella typhimurium* that has a specific frameshift or base pair mutation that makes it His$^-$ to chemicals. Those chemicals that can mutate the bacteria to His$^+$ are mutagenic. Some chemicals become mutagenic only when a liver extract, S-9, is added to the medium. This extract oxidizes the chemicals to a mutagenic form much as the mammalian system inadvertently does. The Ames test has proven to be a useful primary test to screen for potential mutagens and carcinogens.

9.7 Gene Transfer

Gene transfer—the movement of genetic information from one bacterium to another. There are three processes for genetic transfer in bacteria: **transformation, transduction**, and **conjugation**. All of these are significant in that they bring about an increase in the amount of genetic variation within a population.

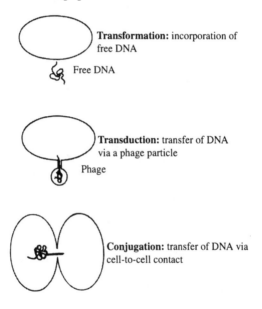

Transformation: incorporation of free DNA

Free DNA

Transduction: transfer of DNA via a phage particle

Phage

Conjugation: transfer of DNA via cell-to-cell contact

Figure 9.3 Summary: Three Methods of Gene Transfer in Bacteria

9.7.1 Transformation

Transformation—"naked" DNA is transferred from one bacterium to another in solution. DNA fragments are released as double-stranded DNA into the medium. **Endonucleases** cut the double-stranded DNA in solution and the resulting fragments separate—only single-stranded molecules are transferred. The transferred DNA is spliced into the recipient cell's DNA. Uptake is dependent upon the presence of a protein known as competence factor. **Bacterial competence** is the ability of a bacterium to take up DNA from the extracellular environment.

Transformation occurs naturally among some bacteria. It is used in the laboratory to create recombinant DNA and is also used to study the effects of introducing DNA into a cell, and in mapping gene locations.

Problem Solving Example:

 What is transformation?

Transformation is a means by which genetic information is passed in bacterial cells. The recipient cell takes up the DNA that has been released by the donor cell. This occurs naturally in some species; however, it is usually performed as part of an experimental procedure. The DNA is extracted from the donor cell and mixed with recipient cells. *Hemophilus influenzae* and *Bacillus subtilis* are naturally competent; they are capable of taking up high molecular weight DNA from the medium. Competent cells have a surface protein called competence factor, which binds DNA to the cell surface. Other cells, such as *E. coli,* cannot readily undergo transformation. They will only pick up extracellular DNA under special laboratory conditions. The cells must have mutations that stop exonuclease I and V activity. The cells must be treated with high $CaCl_2$ concentrations to make their membranes permeable to the DNA. The donor DNA must be present in very high concentrations.

The DNA that is picked up by the recipient cell must be double-stranded. As it enters the cell, an intracellular DNAase degrades one of the strands. This hydrolysis provides the energy needed to pull the rest of the DNA into the cell. Once inside the cell, the now single-stranded DNA can insert into homologous regions of the recipient's chromosome. When the donor DNA and recipient DNA have genetic mutations that act as markers, genetic linkage can be established through transformation experiments.

9.7.1.1 The Experiments of Griffith and Avery

The following experiment of Frederick Griffith (1928) demonstrated transformation. There are two kinds of pneumococcus cells—rough and smooth. Only the smooth form is **virulent** (i.e., capable of infecting and killing mice). Griffith showed that neither heat-killed smooth cells nor live rough cells alone were capable of causing infection. However, when live rough cells were mixed with heat-killed smooth cells, the mice died and live smooth cells could be recovered. Oswald T. Avery (1940) showed that only the smooth cells contained the capsular polysaccharide that is responsible for virulence, and that the substance involved in the transfer was neither a protein nor the polysaccharide itself. Instead, it was the DNA containing the gene for the capsular polysaccharide that was being transferred from the dead smooth cells to the live rough cells, transforming them into the virulent smooth pneumococci.

Problem Solving Example:

Q What is the nature of a "transforming agent"? What importance may this phenomenon have on our understanding of the chemical basis of inheritance?

A The concept of transformation arises from the experiments performed by Griffith in 1928. It was observed that when injected into mice, some strains of pneumococcus bacteria caused pneumonia and usually death. Other strains of the bacteria were relatively harmless. The infective form always had a capsule (a complicated polysaccharide coating). The noninfective form did not have a virulent capsule. The encapsulated strain was called the "smooth strain" because the colonies looked smooth on a culture plate, and the harmless, unencapsulated strain was called the "rough strain" because of the rough appearance of its colonies.

In his famous experiment, Griffith injected one group of mice with the virulent smooth strain and another with the harmless rough strain. As expected, mice from the former group died, while the latter group survived. Griffith then injected a third group of mice with the heat-killed smooth strain. This group lived, showing that bacteria killed by heat were no longer virulent. However, when a fourth group of mice was injected simultaneously with both the harmless rough strain and the heat-killed smooth bacteria, the mice died. The disease-causing organisms were of the smooth type bacteria. This process, by which something from the heat-killed bacteria converted rough bacteria into smooth bacteria, is known as *transformation.*

Griffith felt that protein from the dead bacteria might be the active transforming agent. But Griffith's interpretation of his experiment was later shown to be incorrect by Avery, MacLeod, and McCarthy in 1944. They made an extract from heat-killed smooth cells and purified the extract by removing any substance that did not cause transformation of rough bacteria into smooth bacteria. Eventually, they determined that DNA was the essential transforming agent. Moreover, when this extracted DNA was added to other types of rough strains, the bacteria formed from transformation were always identical to the bacteria that donated the DNA. This indicated that hereditary information must be carried by DNA.

Transformation suggests that in bacteria, genetic traits can be passed via DNA alone, from one bacterium to another.

9.7.2 Transduction

Transduction—DNA is transferred from one bacterium to another via a bacteriophage and then incorporated into the recipient's DNA. The bacterial virus may be virulent or temperate (i.e., a prophage).

Problem Solving Example:

 What is transduction?

Transduction is a phage-mediated transfer of genetic material between two bacteria. There are three types of transduction: generalized, specialized, and F-mediated.

Generalized transduction occurs when a piece of bacterial chromosome becomes incorporated in a phage head and is transferred to a recipient bacteria. It can become incorporated in the recipient by recombination at the homologous region of the chromosomal DNA. The bacterial DNA becomes incorporated at the end of the phage's lytic cycle. The phages that can accomplish this are not as selective towards the DNA fragments that they incorporate as those that cannot transduce. But the incorporation of bacterial DNA is not always so simple. For example, in phage P22 a gene can recognize particular signals in the *Salmonella* chromosome. The gene is responsible for cutting up the DNA into sizes appropriate for the head protein. The signals that the gene recognizes may be certain base sequences and, as a result, certain *Salmonella* markers are transduced more frequently than others.

Specialized transduction results in restricted parts of the bacterial chromosome being incorporated into the phage particle. This happens when the original transducing particle is produced by a faulty outlooping of the prophage. The phage thus formed is defective since some of its genes remain in the bacteria replaced by some of the bacterial genes.

F-mediated sexduction is the third form of transduction. An F element that contains extra bacterial genes is called an F' element. By conjugation, the F' element can be transferred to a bacterial cell that is mutant in the extra genes on the F' element. Cells are then selected for those that have the F' element integrated next to where a phage has integrated. When the prophage is induced, the phage that results will probably contain the genes from the F' element. This virus is then used as the vector to transfer the bacterial genes to another bacterial cell.

These viral-mediated gene transfer mechanisms have been exploited by geneticists to map bacterial chromosomes. Very detailed maps that are continuously being revised are the result.

9.7.3 Conjugation

Conjugation—DNA is transferred from one live bacterium to another through direct contact; large quantities of DNA can be transferred in this way.

F factors—plasmids transferred from a donor cell (an F^+ cell) to a recipient cell (an F^- cell) during conjugation.

An **Hfr (high frequency of recombination)** is a cell with an F plasmid incorporated into the chromosome.

Problem Solving Example:

Q A male bacterium conjugates with a female bacterium. After conjugation, the female becomes a male. Account for this "sex change."

A Conjugation occurs between bacterial cells of different mating types. Maleness in bacteria is determined by the presence of a small, extra piece of DNA, the sex factor, which can replicate itself and exist autonomously (independent of the larger chromosome) in the cytoplasm. Male bacteria having the sex factor, also known as the F factor, are termed F⁺ if the sex factor exists extrachromosomally. F⁺ bacteria can only conjugate with F⁻, the female counterparts, which do not possess the F factor. Genes on the F factor determine the formation of hairlike projections on the surface of the F⁺ bacterium, called F or sex pili. The pili form cytoplasmic bridges through which genetic material is transferred and aids the male bacterium in adhering to the female during conjugation. During conjugation of an F⁺ with an F⁻ bacterium, the DNA that is the most likely to be transferred to the female is the F factor. Prior to transfer, the F factor undergoes replication. The female thus becomes a male by receiving one copy of the F factor, and the male retains its sex by holding on to the other copy. The DNA of the male chromosome is very rarely transferred in this type of conjugation.

If this were the only type of genetic exchange in conjugation, all bacteria would become males and conjugation would cease. However, in F⁺ bacterial cultures, a few bacteria can be isolated that have the F factor incorporated into their chromosomes. These male bacteria that conjugate with F⁻ cells are called Hfr (high frequency of recombination) bacteria. They do not transfer the F factor to the female cells during conjugation, but they frequently transfer portions of their chromosomes. This process is unidirectional, and no genetic material from the F⁻ cell is transferred to the Hfr cell.

9.8   Recombination

Genetic recombination—rearrangement of genes from separate groups of genes. Genes from two chromosomes are recombined into one chromosome containing some genes from each of the original chromosomes. This contributes to genetic diversity. In eukaryotes, the process is associated with sexual reproduction, during which haploid gametes, produced by meiosis, fuse to form a diploid zygote. Portions of chromosomes may be exchanged during meiosis by a process known as crossing-over.

Problem Solving Example:

Q A bacterial strain is unable to synthesize the amino acids methionine and histidine and is also unable to ferment arabinose. It is transduced by a phage with the wild-type genome, met^+ his^+ ara^+. Recombinants are selected for by growth on plates supplemented with histidine. The colonies that grew on his^+ plates were placed on plates containing arabinose. A total of 320 colonies grew on histidine-supplemented plates, and 150 of these could also ferment arabinose. What is the amount of recombination between met and ara?

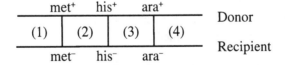

A Transduction is the process whereby bacterial DNA is transferred from one cell to another via a phage vector. The phenotype of the recipient bacteria will be altered to that of the donor if recombination occurs between the incoming and the native DNA.

Crossing-over in the different regions produces different recombinants. Since the original plates were supplemented with histidine, both his⁺ and his⁻ cells will be able to grow; crossing-over in regions (2) and (3) are thus not accounted for in the selected colonies. Crossing-over in regions (1) and either (2) or (3) produces prototrophic mutants (met⁺ara⁻) that can grow on unsupplemented medium. The frequency with which recombination occurs between met and ara can be found by dividing the number of met⁺ara⁻ prototrophs by the total number of recombinants.

recombination ratio = # of met⁺ara⁻
$$\qquad\qquad\qquad \text{# of met}^+\text{ara}^+$$

$$= \frac{320 - 150}{320}$$

$$= 0.531 \text{ or } 53.1 \text{ percent}$$

9.9 Transposons

Transposons—small DNA segments that can move from one part of a chromosome to another area on either the same chromosome, a different chromosome, or a plasmid. They may be **simple** or **complex**. A simple transposon, or insertion sequence, is a short segment of DNA that contains only those genes coding for the enzymes responsible for its transposition. Complex transposons can carry any type of gene, including those for antibiotic resistance, and are an important natural mechanism for gene movement.

Problem Solving Example:

Q Describe some of the genes that can be transferred to a host genome via a transposon.

A Transposons, or "jumping genes," were discovered by Barbara McClintock in maize. They are DNA segments measuring from 700 to 40,000 bp in length that can move from one region of the genome to another. A transposon contains the enzyme transposase, which is responsible for the transposition of DNA segments and for recognition sites, which are short inverted repeats. Transposons can also contain other genetic material, including antibiotic resistance genes and genes that encode toxins.

9.10 Recombinant DNA Technology

Genetic engineering—manipulated gene transfer in the laboratory.

Recombinant DNA—DNA that has been artificially altered to combine genes from different sources.

Biotechnology—application of genetic engineering and use of recombinant DNA in research, medicine, industry, and agriculture (see also Chapter 12).

Vector—piece of DNA used to transfer DNA between organisms. The gene of interest is inserted into the vector (usually a plasmid or a viral genome), which is then transferred into a new cell. The new cell is used to grow a **clone** from which large amounts of the gene or its product can be harvested. Some vectors contain antibiotic-resistance genes, or **markers**, that can be used to identify cells containing the vector.

Restriction enzyme—an enzyme that recognizes and cuts a specific DNA sequence. They are generally named for the microorganism in which they were first discovered; e.g., the restriction enzyme *EcoRI* was isolated from *Escherichia coli;* it recognizes and cuts the sequence GAATTC.

Gene library—an entire genome cut with restriction enzymes and inserted into vector molecules.

cDNA—DNA synthesized from a strand of mRNA by reverse transcription; the required enzyme is **reverse transcriptase**, which was originally isolated from a retrovirus.

Clone identification can be accomplished through replica plating.

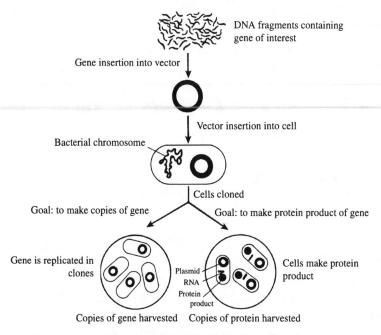

DNA fragments containing gene of interest

Gene insertion into vector

Vector insertion into cell

Bacterial chromosome

Cells cloned

Goal: to make copies of gene

Goal: to make protein product of gene

Gene is replicated in clones

Plasmid
RNA
Protein product

Cells make protein product

Copies of gene harvested   Copies of protein harvested

Figure 9.4 Genetic Engineering

9.11 Diversity and Evolution

Mutation, gene transfer, and recombination are all important in increasing genetic diversity. **Diversity** is a necessary prerequisite for evolution.

Problem Solving Example:

Q Genetic variation occurs in bacteria as a result of mutations. This would seem to be the only process giving rise to variation, for bacteria reproduce through binary fission. Design an experiment to support the existence of another process that results in variation.

A   It was originally thought that all bacterial cells arose from other cells by binary fission, which is the simple division of a parent bacterium. The two daughter cells are genetically identical because the parental chromosome is simply replicated, with each cell getting a copy. Any genetic variation was thought to occur solely from mutations. However, it can be shown that genetic variation also results from a mating process in which genetic information is exchanged.

One can show this recombination of genetic traits by using two mutant strains of *E. coli,* which lack the ability to synthesize two amino acids. One mutant strain is unable to synthesize amino acids A and B, while the other strain is unable to synthesize amino acids C and D. Both these mutant strains can be grown only on nutrient media, which contain all essential amino acids. When both strains are plated on a selective medium lacking all four amino acids in question, some colonies of prototrophic cells appear that can synthesize all four amino acids (A, B, C, and D). When the two strains are plated on separate minimal medium plates, no recombination occurs, and no prototrophic colonies appear.

These results may be proposed to be actually a spontaneous reversion of the mutations back to wild type (normal prototroph) rather than ecombination. If a single mutation reverts to wild type at a frequency f 10^{-6} mutations per cell per generation, two mutations would simultaneously revert at a frequency of $10^{-6} \times 10^{-6} = 10^{-12}$ mutations per cell per generation. If one plates about 10^9 bacteria, no mutational revertants for both deficiencies should occur. Recombination occurs at a frequency of 10^{-7}, and thus if 10^9 bacteria are plated, prototrophic colonies should be found.

This recombination of traits from the two parent mutant strains is brought about through a process called conjugation. During conjugation, two bacterial cells lie close to one another and a cytoplasmic bridge forms between them. Parts of one bacterium's chromosome are transferred through this tube to the recipient bacterium. The transferred chromosomal piece may or may not get incorporated into the recipient bacterium's chromosome.

9.12 Regulation of Gene Expression in Bacteria

Constitutive enzymes—those that are always present. **Constitutive genes** are those that continue to produce proteins regardless of other factors, including the concentrations of substrate and end product. For example, most glycolytic enzymes (and the genes that produce them) are constitutive.

Other enzymes are regulated at the genetic level and are described as **inducible**. Regulatory mechanisms such as **induction**, **repression**, and **attenuation** are ways to control the activity of genes to determine which mRNAs are synthesized and, therefore, which proteins will be made.

9.12.1 The Operon Model

An **operon** consists of three segments: a promoter, an operator, and structural genes. When the repressor, which is the protein that binds to the operator, is removed, transcription can take place. Gene expression occurs once transcription of the structural genes is set in motion.

9.12.2 Enzyme Induction

In the **operon model for an inducible system**, the presence of an inducer activates an operon and the cell synthesizes more enzymes. In some cases, the inducer binds to the repressor so that it cannot bind to the operator and prevent transcription.

Beta-galactosidase and the lac operon is an example of an inducible system. Beta-galactosidase is an enzyme involved in the metabolism of lactose. Lactose is the inducer for the operon controlling production of beta-galactosidase. When lactose is absent, a repressor is produced that binds to the operator and inactivates the operon—beta-galactosidase is not produced. When lactose is present, it inactivates the repressor, thus allowing transcription to occur and beta-galactosidase is produced.

9.12.3 Enzyme Repression

In the **operon model for a repressible system**, a repressor binds to the operator and prevents transcription; in some cases the repressor cannot bind to the operator site without a **corepressor,** and it is the presence or absence of the corepressor that controls synthesis.

The repressor is often a synthetic product that inhibits further production of the enzyme responsible for its synthesis. **Tryptophan and the trp operon** illustrate enzyme repression. When tryptophan is present, it attaches to and activates a regulator protein that represses the trp operon. When tryptophan is not present, the repressor is not activated and transcription of the trp operon can occur.

Catabolite repression—the presence of glucose (or some other preferred nutrient) represses synthesis of the enzymes necessary to metabolize an alternative substance.

9.12.4 Attenuation

Attenuation is a form of regulation wherein mRNA synthesis is prematurely terminated at a point called the **attenuator** site.

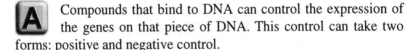

Problem Solving Examples:

 What is the difference between negative and positive control of gene expression?

Compounds that bind to DNA can control the expression of the genes on that piece of DNA. This control can take two forms: positive and negative control.

Negative control is operative in some prokaryotic catabolic systems (see the first figure). The *lac* operon of *E. coli* is under negative control. This form of control utilizes repressors and inducers to turn off a genetic system that would otherwise be turned on. The repressor molecule interacts with the DNA to inhibit the synthesis of the gene products. This inhibition is terminated when the inducer, the molecule that is to be catabolized, is present. Thus, negative control involves substances that inhibit gene activity.

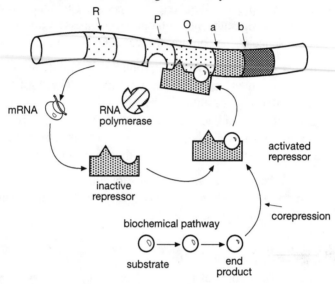

Positive control is found in biosynthetic as well as catabolic systems. The *lac* operon is under positive control as well as negative control. Positive control occurs when components enhance gene activity. Hormones, special proteins such as the catabolite activator protein (CAP), and cyclic AMP can act to enhance the transcription of genes (see the figure below).

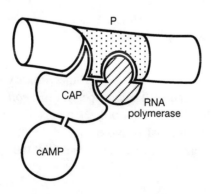

Genes can be controlled by positive and/or negative control. Catabolic systems, such as the breakdown of lactose, utilize negative control since the genes need to be expressed only in the presence of the compound to be degraded. Biosynthetic systems, such as tryptophan biosynthesis, use positive control since they synthesize a product. There is still much to learn about the regulation of gene activity. But it seems that different forms of regulation are needed for different types of gene expression.

 Explain how gene expression is regulated during phage λ infection.

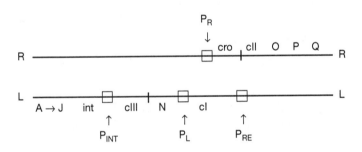

 The phage λ has a system of operons and regulatory proteins that regulate whether it will undergo lysis or lysogeny when it infects a host. Whether the virus undergoes lysis or lysogeny depends on environmental stimuli which affect gene expression. The genes that are involved in this regulation are ordered strategically on the λ chromosome. Contrary to most chromosomes, each strand of DNA is used as a sense strand—one is read towards the right and the other is transcribed towards the left. A linear representation of the λ chromosome is shown in the figure above.

There are three phases that regulate λ's lytic cycle. They are the immediate-early, delayed-early, and late-phase. The immediate-early phase begins immediately after the phage infects a cell. The host's polymerases bind to a promoter region (P_R) of the phage genome. P_R directs transcription on the R strand in the rightward direction. The second immediate-early transcript begins at the leftward promoter (P_L), which transcribes to the left. The transcription that begins at P_R leads to the production of Cro protein, and the transcription that begins at P_L leads to the production of N protein. N protein is an antiterminator. It enables the RNA polymerase to get over the chain terminating sequences that are denoted by black bars in the first figure. N protein also stimulates the transcription of the delayed-early phase. The delayed-early genes are cII and cIII, which are used only in lysogeny; genes O and P, which are needed for the replication of the phage chromosome; and gene Q, which is needed for the stimulation of the next phase. When the Q gene has been transcribed, the Q protein, another antiterminator, stimulates the late-phase genes. These genes, denoted A $\rightarrow$ J in the first figure, encode the head and tail proteins of the phage. Once these proteins are made, the replicated phage chromosomes are packaged inside protein coats and the host cell is lysed. The figure below shows the three lytic phases. The genome is actually circular, so the third phase is not actually split as it appears in the diagram.

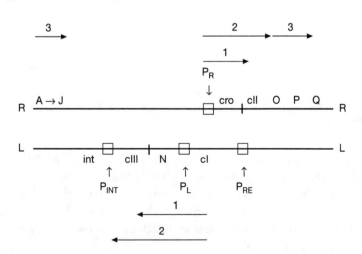

Lysogeny requires two events: the repressor protein, cI, must be synthesized to turn off the lytic cycle by binding to operator regions near the promoters, P_R and P_L, and the Int protein must be synthesized to mediate λ's integration into the bacterial host's genome. In order for cI to be transcribed, a polymerase must bind to the P_{RE} promoter. P_{INT} must be recognized by a polymerase for the Int protein to be transcribed. Both promoters are recognized by RNA polymerase only in the presence of cII and cIII proteins. The cII protein acts as an activator protein; it stimulates polymerase to bind to promoters.

The race between lysis and lysogeny pivots on the levels of Cro protein and cI. Cro protein can act as a repressor by binding to the operators that cI can bind. It is not as stable as cI, nor does it bind as strongly. Thus, when it binds it slows down, but does not stop, the expression of N, cIII, Cro, cII, O, P, and Q genes. When there is a lot of Cro protein, little cII and cIII is transcribed, and thus no cI is made. When there is no cI, there is no strong repression of the lytic cycle phases, so the phage lyses the cell. When there is a small amount of Cro, the cII and cIII levels are high because Cro is not slowing their synthesis. These proteins are necessary for cI transcription, so cI is produced. The cI protein strongly inhibits the lytic cycle genes, and the phage undergoes lysogeny.

The state of the phage is influenced by the levels of the proteins that are produced. Cro protein levels are affected by temperature, the metabolic state of the host, the genotype of the host, and the genotype of the infecting phage. The lytic cycle will be followed when the conditions are optimal for the survival of the phage's progeny. Otherwise, the lysogenic cycle will be followed.

Quiz: Microbial Genetics

1. Certain oncogenic viruses violate the central dogma of biochemical genetics by transferring genetic information in which one of the following sequences?

    (A) DNA, protein, RNA

    (B) DNA, RNA, protein

    (C) RNA, DNA, protein

    (D) Protein, DNA, RNA

    (E) RNA, protein, DNA

2.   Which stimulus will activate the lactose operon in a bacterial cell?

    (A) Absence of lactose

    (B) Availability of inducer

    (C) Cistron repression

    (D) Regulator gene dominance

    (E) Repressor molecule binding to the operator gene

3.   Restriction enzymes are used in genetic research to

    (A) cleave DNA molecules at certain sites.

    (B) produce individual nucleotides from DNA.

    (C) slow down the reproductive rate of bacteria.

    (D) remove DNA strands from the nucleus.

    (E) prevent histones from reassociating with DNA.

4.   The protein binding and blocking the operator gene in the operon model of gene control is the

    (A) inactivator.

    (B) operator.

    (C) promoter.

    (D) regulator.

    (E) repressor.

5. A bacterial gene 450 bases long will produce a protein containing approximately how many amino acids?

    (A) 150

    (B) 300

    (C) 450

    (D) 900

    (E) 1,350

6. Bacteria contain

    (A) true nuclei.

    (B) chloroplasts.

    (C) mitochondria.

    (D) simple, circular DNA.

    (E) cell membranes as their outermost structures.

7. To find the genetic order of three bacterial genes, we do NOT need to know the

    (A) number of wild-type cells.

    (B) frequency of recombination.

    (C) dominance and recessiveness of the alleles.

    (D) number of double crossover events.

    (E) phenotypes of the cells.

8. During conjugation, what is transferred from the Hfr bacterium to the F⁻ bacterium?

    (A) The sex factor (F factor)

    (B) Portions of the Hfr chromosome

    (C) The sex factor and portions of the Hfr chromosome

(D) Nothing is transferred.

(E) None of the above.

9. Sometimes it is found that viruses can transfer genetic material from one bacterial strain to another. This process is called

(A) transduction.

(B) recombination.

(C) conjugation.

(D) transmission.

(E) mutation.

10. The direct transfer of bacterial DNA from one organism to another is

(A) transcription.

(B) translation.

(C) transformation.

(D) translocation.

(E) transduction.

ANSWER KEY

1.	(C)	6.	(D)
2.	(B)	7.	(C)
3.	(A)	8.	(B)
4.	(E)	9.	(A)
5.	(A)	10.	(C)

CHAPTER **10**

Role of Microbes in Disease

10.1 Host–Microbe Relationships

An organism that harbors another organism is called a **host.**

Normal (indigenous) flora—those microorganisms that can establish populations in a host, such as the human body, without causing disease. These microorganisms are called normal (indigenous) flora.

Opportunistic flora are those microorganisms that do not normally cause disease but may do so under certain conditions (e.g., when the host is immunocompromised); they may be resident or transient.

Resident flora are those with permanent populations.

Transient flora are those with temporary or semipermanent populations.

Microbial antagonism is a phenomenon wherein the normal flora prevent pathogens from causing infection.

Problem Solving Example:

 Why are parasites usually restricted to one or a very few host species?

 Parasites usually have extremely specific requirements for growth. They may grow only within a narrow range of pH and temperature, at a certain oxygen concentration, and may require a large number of different organic nutrients. The specific combination of optimal growing conditions that is required by a particular parasite can be found only in one host species, or in several closely related host species. Usually the parasite can live only in certain locations within the host. The tapeworm, for example, can live only in the intestine of the human and will not infect the kidneys or the bones. Saprophytes, on the other hand, will grow within a broad range of temperatures and pH and O_2 concentrations, and they require very few organic nutrients. Yeasts, an example of a saprophyte, are able to grow at many different oxygen concentrations, oxidizing glucose to CO_2 and water if oxygen is present, and fermenting glucose to CO_2 and ethanol if oxygen is not present in sufficient amounts. Parasites, in general, are unable to switch from oxidative to fermentative metabolism and must utilize either one or the other. In addition, yeasts can synthesize all their constituent proteins, nucleic acids, and other components if they are supplied with glucose. Parasitic bacteria lack the enzymes needed for this and must be supplied with amino acids, vitamins, and a mixture of sugars.

These complex growth requirements make it difficult to culture parasitic organisms in the laboratory. Many parasites can grow only if supplied with extracts from animal tissues. Some parasites, such as viruses and rickettsias, can grow only in the presence of living cells.

10.1.1 Symbiosis: Commensalism, Mutualism, and Parasitism

A relationship wherein two organisms live together is called a **symbiosis**; there are three types:

Commensalism—one of the two organisms benefits from the relationship, the other is unaffected.

Mutualism—both organisms benefit from the relationship.

Parasitism—one organism benefits and the other is harmed.

Problem Solving Examples:

 What is mutualism? Give three examples of this type of relationship.

 Mutualism, like parasitism and commensalism, occurs widely through most of the principal plant and animal groups and includes an astonishing diversity of physiological and behavioral adaptations. In the mutualistic type of relationship, both species benefit from each other. Some of the most advanced and ecologically important examples occur among the plants. Nitrogen-fixing bacteria of the genus *Rhizobium* live in special nodules in the roots of legumes. In exchange for protection and shelter, the bacteria provide the legumes with substantial amounts of nitrates which aid in their growth.

In humans, certain bacteria that synthesize vitamin K live mutualistically in the human intestine, which provides them with nutrients and a favorable environment.

 In a commensalistic relationship between two species living symbiotically, what are the consequences of their interactions?

 Commensalism is a relationship between two species in which one species benefits, while the other receives neither benefit nor harm. The advantage derived by the commensal species from its association with the host frequently involves shelter, support, transport, food, or a combination of these.

10.1.2 Pathogens

Pathogen—a parasitic microorganism that causes disease.

Pathogenicity—the ability of a microbe to produce disease.

Virulence—the power of an organism to cause disease.

Infection—pathogen invasion of the body.

Disease—disturbed health due to a pathogen or other factor.

Pathogens gain access to a host via a portal of entry and leave via a portal of exit.

Portals of entry include the mucous membranes (including those lining the respiratory, gastrointestinal, and genitourinary tracts) and parenteral entry (direct inoculation through the skin via bites, injections, or other wounds). The most frequently used portal of entry is the respiratory tract. Many pathogens cannot cause infection unless they enter through a specific (**preferred**) portal of entry.

Portals of exit include the respiratory tract (through coughing and sneezing), the gastrointestinal tract (through saliva and feces), and secretions from the genitourinary tract. Microbes may also leave the body via blood into syringes or biting arthropods.

Problem Solving Examples:

 Distinguish between infection and infestation and between virulence and pathogenicity.

 Although these terms are used interchangeably, the meanings of these words are rather distinct.

The term infection implies an interaction between two living organisms. The host and the parasitic microorganism compete for superiority over the other. If the microorganism prevails, disease results. If the host is dominant, immunity or increased resistance to the disease may develop.

Infestation indicates the presence of animal (nonmicrobial) parasites in or on the host's body. Lice, fleas, and flatworms are infesting organisms. They may transmit an infection (e.g., a louse carries typhus—a disease caused by microorganisms called rickettsia) to humans.

Parasitism is a type of antagonism in which one organism, the parasite, lives at the expense of the other—the host. Infection is a type of parasitism.

Pathogenicity refers to the ability of a parasite to gain entrance to a host and produce disease. (A pathogen is any organism capable of producing disease.) The degree of pathogenicity, or ability to cause infection, is called virulence. The virulence of a microorganism is not only determined by its inherent properties but also by the host's ability to resist the infection. A pathogen may be virulent for one host and nonvirulent for another. For example, streptococci, although found in the throats of some healthy individuals, can be pathogenic under different conditions or in other individuals.

Resistance is the ability of an organism to repel infection. Immunity is resistance (usually to one type of microorganism) developed through exposure to the pathogen by natural or artificial means. Lack of resistance is called susceptibility.

 What are the bacterial factors influencing virulence?

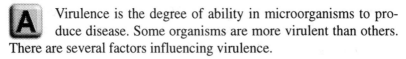 Virulence is the degree of ability in microorganisms to produce disease. Some organisms are more virulent than others. There are several factors influencing virulence.

One of these is the production of toxins, which are poisonous substances produced by some microorganisms. Both the ability to produce the toxin and the potency of the toxin affect the organism's capability to produce disease. Toxins that are secreted into the surrounding medium during cellular growth are called exotoxins, and toxins retained in the cell during growth and released upon cell death and lysis are called endotoxins. Exotoxins are released into the surrounding medium, e.g., into a can of vegetables containing *Clostridium botulinum* (the bacterium causing botulism) or into damaged tissue infected with *Clostridium tetani* (the bacterium causing tetanus). Both these types of bacteria exhibit difficulty in penetrating the host. However, these bacteria are virulent because of their toxins. When the spatial configuration of the amino acids in a toxin molecule is altered, the toxicity is lost, and the resulting substance is called a toxoid. Both toxins and toxoids are able to stimulate the production of antitoxins, substances made by the host that are capable of neutralizing the toxin. Endotoxins are liberated only when the microorganisms disintegrate and are generally less toxic than exotoxins. They do not form toxoids and are usually pyrogenic, inducing fever in the host (some exotoxins produce fever also). Exotoxins are usually associated with gram-positive bacteria, while endotoxins are associated with the gram-negative ones.

Another factor influencing virulence is the ability of bacteria to enter the host and penetrate host tissue. Specific bacterial enzymes are involved. One enzyme that is produced by some of the clostridia and cocci is hyaluronidase. This enzyme facilitates the spread of the pathogen by aiding its penetrance into host tissues. It hydrolyzes hyaluronic acid, an essential tissue "cement," and thus increases the permeability of tissue spaces to both the pathogen and its toxic products.

The virulence of pathogens is also influenced by their ability to resist destruction by the host. In certain bacteria, this resistance is due to the presence of a nontoxic polysaccharide material, which forms a capsule surrounding the bacterial cell. For example, pneumococci are *Streptococcus pneumoniae* which are virulent when capsulated but are often avirulent when not capsulated. Capsulated pneumococci are resistant to phagocytosis (the ingestion and destruction of microorganisms by host phagocytes), while those without capsules are ingested by leukocytes and destroyed.

10.2 Kinds of Disease

An **infectious disease** is one caused by an organism that can be transmitted.

A **communicable (contagious) disease** can be spread from one organism (host) to another. A noncommunicable disease cannot be spread in this way and is usually contracted from an environmental source, such as soil or water.

A **primary infection** occurs in a previously noninfected individual.

Secondary infections are those that occur once a host is weakened from a **primary** infection.

A **systemic infection** is spread throughout the body, while a **local infection** is limited to a small area of the body.

An **endogenous disease** is one caused by a pathogen, usually an opportunistic organism, from within the body, while an **exogenous disease** is one caused by a pathogen or other factors from outside the body.

A **chronic disease** is one in which symptoms are slow to develop and the disease is slow to disappear. An **acute disease** progresses quickly.

Latent symptoms appear (or reappear) long after infection.

An infection that does not cause any signs of disease is said to be **inapparent** or **subclinical.**

A **mixed infection** is one caused by two or more pathogens.

Syndrome—a characteristic group of signs and symptoms that always accompanies a specific disease.

Problem Solving Example:

 "A virulent organism is as good as dead if it is not communicable." Explain.

A virulent pathogen will eventually die if it is restricted to its original host. The host produces substances that either prevent the growth and spread of the pathogen or actually destroy it. Even if the host fails to do this, a virulent pathogen will bring about its own destruction by killing the host that sustains it. For the pathogen to survive and cause disease in a number of organisms, it must find new hosts to infect.

Communicable pathogens are transferred from one host to another by a number of means. Pathogens of the respiratory tract, such as *Diplococcus pneumoniae,* can leave the body in discharges from the mouth, nose, and throat. Sneezing and coughing expedite the spread of these organisms. Enteric (intestinal) pathogens, such as those that cause typhoid fever, leave the host in fecal excretions and sometimes in the urine. The bacteria in these wastes contaminate the food and water that a subsequent host may ingest. Other pathogens cannot live long outside a host and are transmitted by direct contact, most commonly through breaks in the skin or contact of mucous membranes. The final common means of transmittance is by certain insect vectors, although these insects may be carried further by other organisms such as rats and humans. For example, ticks transmit *Rickettsiae ricketsii* to humans, causing Rocky Mountain spotted fever.

Some pathogens are transmitted by specific means and enter the new host by a specific portal of entry. For example, enteric bacteria have a special affinity for the alimentary tract and are able to survive both the enzymatic activity of the digestive juices and the acidity of the stomach.

We thus see that the success of a pathogen depends on its successful transmittance to a new host via air, food, water, insects, or by contact.

10.3 How Microbes Cause Disease

Bacterial pathogens must first adhere to a host. **Adhesins** are projections on the surface of the bacterium that adhere to complementary receptors on host cells. **Adherence** is followed by **colonization** of the tissues (in complex multicellular organisms), and may also involve **invasion** of cells.

Disease-causing bacteria may produce toxins or special enzymes. **Hemolysins,** which destroy red blood cells, and **leukocidins,** which destroy neutrophils and macrophages, are both special enzymes produced by pathogenic bacteria. Other such enzymes include **coagulase**, which helps in blood clotting; **fibrinolysin**, which breaks down blood clots; and **hyaluronidase**, which destroys a mucopolysaccharide that holds cells together in tissues.

Staphylokinase is produced by *Staphylococcus aureus*. **Bacterial kinases** are used to dissolve clots in plasma, which is the host cell's attempt to isolate the infection.

Collagenase is produced by *Clostridium* (spp) and breaks down the connective tissue of the host, allowing the pathogen to spread.

M protein is produced by *Streptococcus pyogenes* and assists the attachment of the bacterium to the host epithelial lining. The M protein increases the virulence of the bacterium.

The **capsule** is an indication of the virulence of the microorganism. It increases the virulence by inhibiting the host's natural defense of phagocytosis.

Necrotizing factors are produced by bacteria and cause the death of somatic cells. Necrotizing factors increase the virulence of the bacterium.

10.3.1 Bacterial Toxins

Bacterial toxins include exotoxins and endotoxins.

10.3.1.1 Exotoxins

Exotoxins are produced and excreted mostly by gram-positive bacteria into the surroundings as the pathogen grows. Exotoxins are proteins that usually have enzymatic activity. There are three types of exotoxins: **cytotoxins**, which cause the death of host cells; **neurotoxins**, which interfere with the normal nervous function; and **enterotoxins**, which affect the intestinal lining of the host.

The host immune system produces **antitoxins** in the form of antibodies against bacterial exotoxins. Exotoxins may also be inactivated by heat or chemical treatment. **Toxoids** are altered exotoxins that allow the host immune system to produce antitoxins without causing any side effects. Toxoids may be used as vaccines against the microorganisms that produce the exotoxins.

Corynebacterium diphtheriae produces **diphtheria toxin,** which is cytotoxic to human cells by causing inhibition of protein synthesis. This toxin contains two polypeptides, one that causes symptoms and one that causes binding of the bacterium to the host cell. Diphtheria can be prevented by using a toxoid as a vaccine.

Streptococcus pyogenes produces three cytotoxic **erythrogenic exotoxins,** which cause the characteristic skin rash in scarlet fever.

Clostridium botulinum produces the neurotoxin **botulinum toxin.** Botulinum toxin binds to acetylcholine, a neurotransmitter, and causes paralysis.

Clostridium tetani produces the neurotoxin **tetanospasmin** or the **tetanus toxin.** This toxin binds specifically to cells that control the contraction of skeletal muscle. The result of toxin binding is uncontrollable muscle contraction (lockjaw).

Vibrio cholerae produces the enterotoxin cholera toxin. The **vibrio enterotoxin** consists of two polypeptides, one involved in the symptoms of the disease and the second involved in binding. The bacteria bind to the host's intestinal epithelium and cause the epithelium to excrete fluids and electrolytes. The result is diarrhea, vomiting, and disturbances of normal muscular contraction.

Staphylococcus aureus produces a **staphylococcal enterotoxin** that causes toxic shock syndrome.

10.3.2 Endotoxins

Endotoxins are produced by gram-negative bacteria and are located in the outer membrane of the bacteria. The lipid component of the **lipopolysaccharides (LPS)** is the endotoxin. Endotoxins are not secreted by the bacterial cells. The bacterial cell must die and the outer membrane be broken down for the endotoxin to be released into the bloodstream. The host's responses to endotoxins include chills, fever, weakness, generalized aches, and, in severe cases, shock and death.

The host macrophage in response to the phagocytosis of the gram-negative bacteria secrete **interleukin-1 (IL-1)**. IL-1 causes the hypothalamus to release prostaglandins and results in fever.

Severe reactions to endotoxins include septic or endotoxic shock. Phagocytosis of the gram-negative bacteria causes the secretion of **tumor necrosis factor (TNF)** or **cachectin** from host phagocytes. The net effect of the secretion of these are increases in tissue fluid loss, loss of blood pressure, and shock.

Pathogenic viruses, fungi, protozoans, and helminths also damage host cells and tissues.

Cytopathic effects (CPEs) are the signs of cell damage due to viral infection.

Viruses also release digestive enzymes and alter host DNA.

Fungal pathogens digest cells; some produce toxins or allergic reactions.

Protozoans and helminths produce symptoms through direct damage to host tissues, release of toxic waste products, or by causing allergic reactions.

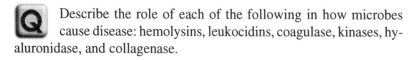

Problem Solving Example:

Q Describe the role of each of the following in how microbes cause disease: hemolysins, leukocidins, coagulase, kinases, hyaluronidase, and collagenase.

A Hemolysins are enzymes produced by bacteria that cause the lysis of red blood cells. For example, *Clostridium perfringens* produces hemolysins and is the most common cause of gas gangrene. Staphylococci and streptococci also produce hemolysins.

Leukocidins are produced by some bacteria and are capable of lysing neutrophils and leukocytes involved in phagocytosis, including macrophages. Staphylococci and streptococci produce leukocidins.

Coagulases are produced by bacteria and are involved in blood clotting. Coagulase converts fibrinogen to fibrin, which forms the clot. The formation of the clot may protect the bacteria from the host immune system. Staphylococci produce coagulases.

Kinases are bacterial enzymes that break down clots by converting plasminogen to plasmin. *Streptococcus pyogenes* produces streptokinase or fibrinolysin, which has been used in treatment of obstruction of the coronary arteries.

Hyaluronidase affects the connective tissue in the host by breaking down the structure of hyaluronic acid. Streptococci and clostridia produce hyaluronidase, which promotes the spread of the infection.

Collagenase is also involved in the spread of infection by breaking down collagen, which forms the framework of connective tissues. *Clostridium perfringens* produces collagenases.

10.4 The Disease Process

Signs of disease are objective changes that can be observed and measured.

Symptoms are subjective changes reported by the patient.

Physicians use both signs and symptoms in making a **diagnosis** (identification of a disease).

Incubation period—the time interval between infection and the appearance of signs and symptoms.

Prodromal period—pathogens are beginning to invade host tissues; characterized by the appearance of early signs and symptoms that are generally nonspecific in nature.

Period of illness—disease is at its most severe; all signs and symptoms are present and at their worst. This is also known as the **invasive** phase. The term **acme**, or **critical**, stage is used to describe the period of most intense symptoms.

Period of decline—signs and symptoms begin to subside; host defenses overcome the pathogens.

Period of convalescence—tissue damage is repaired and the body returns to its healthy, prediseased state.

Problem Solving Example:

Q Distinguish between symptoms and signs in the infectious disease process.

A Symptoms are subjective evidence of disease such as pain and discomfort. These changes cannot be observed but must be reported to the physician. Signs are objective changes that can be observed and measured by a second party, in this case a physician. Examples of signs of disease are swellings, fever, and paralysis. A syndrome is a group of characteristic signs and symptoms of a disease.

10.5 Koch's Postulates

Koch's postulates provide a method for demonstrating that a specific microorganism transmits a specific disease.

Koch's postulates are as follows:

1. The microorganism must be obtained from a diseased animal (the microbe must be found in every animal that has the disease).

2. The microbe is isolated and grown in pure culture.

3. The cultured organism is inoculated into a healthy animal, which then contracts the same disease.

4. Identical microorganisms are recovered from the diseased animal and re-isolated in pure culture.

10.5.1 Exceptions to Koch's Postulates

Some microbial pathogens cause more than one disease, e.g., *Streptococcus pyogenes*.

Some diseases may be caused by more than one microbial pathogen, e.g., pneumonia.

Some diseases do not always exhibit the same signs and symptoms, e.g., tetanus.

Some pathogens cannot be grown on artificial media.

Problem Solving Example:

 Koch's postulates do not apply to all infectious diseases. Why?

Some microorganisms cannot grow on artificial media. Alternative methods for culturing these microorganisms have been used. For instance, *Treponema pallidum,* which causes syphilis, has not been cultured on artificial media. *Mycobacterium leprae,* which causes leprosy, cannot be grown on artificial media.

In addition, some microorganisms may cause several different disease conditions. For example, *Streptococcus pyogenes* can cause sore throat, scarlet fever, skin infections, puerperal fever, osteomyelitis, and others.

10.6 Epidemiology

Epidemiology is the study of the factors and mechanisms involved in the spread of infectious disease; it deals with transmission, incidence, prevalence, and frequency of disease.

Transmission—how a disease is spread among hosts.

Incidence—the number of *new* cases of a disease over a specific period of time.

Prevalence—the number of cases of a disease at a given point in time.

Morbidity rate—the number of cases of the disease, expressed as a proportion of the population.

Mortality rate—the number of deaths attributed to the disease, expressed as a proportion of the population.

Frequency of occurrence may be described as sporadic, endemic, epidemic, or pandemic.

Sporadic—a few isolated cases exist in the population.

Endemic—a large number of cases exist, but do not appear to pose a significant health threat.

Epidemic—a large number of cases exist and are causing patients sufficient harm so as to constitute a significant threat to public health. Epidemics may stem from a **common source** of infection, such as a water supply, or they may be **propagated** through person-to-person contact.

Pandemic—an epidemic disease occurring over an exceptionally large geographic area or areas.

Data on the incidence and prevalence of **reportable** infectious diseases are reported to public health officials on local, state, federal, and world levels. The Centers for Disease Control (CDC), the main source of epidemiological information in the United States, publishes such information in its *Morbidity and Mortality Weekly Report* (MMWR).

Problem Solving Example:

 Discuss the etiology, epidemiology, and prophylaxis of tuberculosis.

It is first necessary to define the terms used in the question before one can answer it. Etiology refers to the cause of a disease. Epidemiology deals with the cause and control of epidemics. An epidemic is the unusual prevalence or sudden appearance of a disease in a community. Prophylaxis is the preventive treatment used to protect against disease.

Many of the important infections of man are airborne and cause diseases of the respiratory tract. Tuberculosis is an endemic respiratory disease; it is peculiar to a locality or a people. Although evidence of its existence has been found in Egyptian mummies 3,000 years old, it is still a leading cause of death. In the nineteenth century, Robert Koch isolated the causative agent of human tuberculosis—*Mycobacterium tuberculosis*. Koch proved that pure cultures of these bacilli would produce the infection in experimental animals, and he later recovered the bacilli from these animals. There are several types of tubercle bacilli (the common name): the human, bovine, avian, and other varieties. One strain is almost exclusively a human parasite, and is responsible for over 90 percent of all cases of tuberculosis. The less common bovine strain can cause tuberculosis in humans through ingestion of infected beef or milk from an infected cow.

It is characteristic of tuberculosis and other respiratory diseases to occur in epidemic form, attacking many people within a short time. Their incidence usually increases during fall and winter, when many people frequently remain indoors.

The tubercle bacillus is transmitted by association with infected people through the secretions of the nose and throat, which is spread in droplets by coughing and sneezing. It may be transmitted by articles that have been used by an infected person, such as eating and drinking utensils and handkerchiefs. The lungs are the most commonly affected tissue, often exhibiting tubercles or small nodules. These are areas of destroyed lung tissue in which the tubercle bacilli grow. Symptoms of tuberculosis include pleurisy (inflammation of the pleural membrane that lines the chest and covers the lungs), chest pains, coughing, fatigue, and loss of weight. Treatment consists of bed rest, nourishing diet, and sometimes chemotherapy. The most effective drug, INH (isonicotinic acid hydrazide), is used both for chemotherapy and chemoprophylaxis. For actual prevention, the drug is administered before infection to highly susceptible individuals.

High death rates in developing countries are due to substandard housing, overcrowding, and malnutrition. Prophylaxis for a community is best achieved when living and working conditions are not overcrowded, diet is adequate, and proper sanitation is available.

10.6.1 Reservoirs

A **reservoir** is a living or nonliving source from which an infectious disease can be spread. Living sources include people who have the disease or are carriers of the disease. Soil, water, and waste materials are all examples of nonliving reservoirs of infection.

Zoonoses are diseases that are transmitted from an animal reservoir to humans. Transmission may be by direct contact or through a vector.

Problem Solving Example:

 Explain how mosquitoes are used as a reservoir for parasites.

Parasites transmitted by mosquitoes are usually carried by their salivary glands. If a mosquito sucks blood containing a parasite, the blood goes to its stomach, where the parasite continues to live. The zygotic stage of the parasite will encyst in the wall of the gut. When the cyst ruptures, the organism migrates to the salivary glands of the mosquito where it is discharged into a new host when the mosquito feeds again.

10.6.2 Transmission

Disease may be transmitted through **direct** or **indirect** contact with a reservoir of infection, or through a vector.

There are two kinds of **direct contact** transmission—**vertical transmission**, i.e., parent to child, and **horizontal transmission**, i.e., person-to-person transmissions other than parent to child.

Indirect contact may occur through **droplets** of saliva or mucus, or through contact with **fomites** (inanimate objects that are contaminated with infectious organisms).

Transmission through a medium such as air, water, or food is known as **vehicle transmission**.

Arthropod vectors (e.g., ticks, fleas, mosquitoes) can transmit pathogens from one host to another by both biological and mechanical means.

Epidemic control measures include elimination of the reservoir, elimination of the vector, immunization of susceptible individuals, and quarantine of infected individuals.

Problem Solving Example:

 Describe four methods of disease transmission.

Contact transmission involves the close proximity between the infectious agent and the susceptible host. Contact transmission can be either direct or indirect. Person-to-person transmission is an example of direct-contact transmission of infectious disease. Diseases that are transmitted in this fashion include the common cold, influenza, staphylococcal infections, hepatitis A, measles, scarlet fever, smallpox, and sexually transmitted diseases.

Indirect transmission occurs when a nonliving object, or fomite, is used to transmit the disease. Examples of indirect transmission include contaminated syringes in the transmission of HIV and hepatitis B. Also, droplet transmission involves the transmission of infectious agents via droplets that travel short distances during coughing, sneezing, laughing, or talking. Influenza, pneumonia, and whooping cough are representative examples.

Vehicle transmission requires a medium such as water, food, air, blood, or body fluids to spread disease. Examples of diseases transmitted by a vehicle are cholera, waterborne shigellosis, and leptospirosis (waterborne); food poisoning and tapeworm (foodborne); and measles, tuberculosis, histoplasmosis, coccidioidomycosis, and blastomycosis (airborne).

Animals or vectors can carry pathogens to susceptible hosts. Arthropods (insects, ticks, mites, fleas) are the most important vectors in transmission of disease. Mechanical transmission is a passive transport with no growth of the organism during transmission. A housefly can transfer pathogens from infected feces to hosts by mechanical (external) means. Examples of mechanical transmission are typhoid fever and shigellosis. Biological transmission in which the organism goes through a morphological or physiological change within the vector is an active process. Malaria and African trypanosomiasis are examples of biological transmission of disease.

10.7 Host Defense Mechanisms

Resistance—the ability to ward off disease. It is the result of genetically predetermined (**innate**) resistance and other factors such as the individual's age, sex, and nutritional status. **Susceptibility** is lack of resistance.

Predisposing factors, such as age, fatigue, stress, and poor nutrition, can make a host more susceptible to infection and disease.

Nonspecific defenses (e.g., fever, inflammation) are used to protect the body from all kinds of pathogenic organisms. They generally serve as a first line of defense.

Specific defenses include **innate resistance** and **acquired resistance** to specific pathogens (**immunity**). Types of immunity will be discussed further in section 10.7.2.

Problem Solving Example:

Define specific defenses, innate resistance, passive immunity, and acquired immunity.

Specific defenses refer to immunity or the characteristic response of the host to a foreign invader or antigen. This response is called the immune response.

Nonspecific immune responses by the host are referred to as innate resistance.

Passive immunity is immunity that is artificially acquired, for instance, the passage of antibodies from breast milk to an infant or artificial antibodies introduced intravenously to the host.

Acquired immunity refers to the body's active response to an antigen and the subsequent production of an immune response. Acquired immunity includes the natural production of the humoral and cell mediated immune responses to an antigen or the targeted production of the immune response to a vaccine.

10.7.1 Nonspecific Host Defense Mechanisms

Nonspecific defenses include **mechanical barriers** such as skin, saliva, the lacrimal apparatus, and mucous membranes, as well as the outward flow of urine, vaginal secretions, and blood (from wounds). **Phagocytosis, fever, inflammation,** and **molecular strategies** are discussed below.

There are three categories of white blood cells (**leukocytes**): the **granulocytes** (neutrophils, basophils, eosinophils), which predominate early in infection; the **monocytes**, which predominate late in infection; and the **lymphocytes.**

Phagocytosis is the cellular ingestion of a foreign substance (including microorganisms). Certain types of white blood cells (including neutrophils and monocytes in the blood, and fixed and wandering macrophages) are **phagocytes**.

Phagocytes locate microorganisms through chemotaxis. They then adhere to the microbial cells, a process that is sometimes facilitated by **opsonization**, wherein the microbial cell is coated with plasma proteins. Pseudopods then encircle and engulf the microbe. The phagocytized microbe, enclosed in a vacuole called a **phagosome**, is usually killed by lysosomal enzymes and oxidizing agents.

Fever is abnormally high body temperature produced in response to infection. It serves to augment the immune system, inhibit microbial growth, increase the rate of chemical reactions, raise the temperature above the organism's optimum growth temperature, and decrease patient activity.

Inflammation is a response to cell damage. Initiation of inflammation is caused by the release of histamine, kinins, and prostaglandins. Redness, heat, swelling, pain, and sometimes loss of function are characteristic of inflammation.

Tissue injury also stimulates **blood clotting**, which may help to prevent dissemination of the infection.

Interferons (see Section 8.2.3) are produced in response to viral infections. They cause uninfected cells to produce **antiviral proteins (AVPs)**.

The **complement system** refers to a group of blood serum proteins that activate a **cascade** series of reactions to destroy invading pathogens. It causes cell lysis, inflammation, and opsonization. Complement deficiencies result in reduced resistance to infection.

Problem Solving Example:

Q If antibodies can be brought into contact with a virus before it attaches itself to the host cell, the virus will be inactivated. Once infection has occurred, the production of interferon may protect the host from extensive tissue destruction. Explain.

A Many human diseases have a viral etiology (cause). Among the more common viral diseases are smallpox, chicken pox, mumps, measles, yellow fever, influenza, rabies, poliomyelitis, viral pneumonia, fever blisters (cold sores), and the common cold. Although most viral infections do not respond to treatment with many of the drugs effective against bacterial infection, many are preventable by means of vaccines.

Buildup of an adequate supply of antibodies requires some time. During this period, extensive viral infection may occur. Most recoveries from viral diseases occur before full development of the immune response (production of antibodies). A factor that is important in the recovery from a viral infection is interferon. Interferon is a protein produced by virus-infected cells that spreads to uninfected cells and makes them more resistant to viral infection. Thus, extensive viral production and resultant tissue damage are prevented.

Upon infection, the host cell is induced to synthesize interferon, a small protein, which is then released into the extracellular fluid and affects the surrounding cells. Interferon binds to surface receptors of uninfected cells. This triggers these cells to synthesize a cytoplasmic enzyme that prevents viral multiplication. Note that the antiviral protein does not prevent entrance of the viral nucleic acid.

Interferon produced by a cell infected with a particular virus can prevent healthy cells from becoming infected by almost all other types of viruses. Interferon is therefore not virus-specific. However, interferon produced by one host species is not effective if administered to a different species. It is therefore host species specific. Since interferon produced by birds and other mammals cannot be used in treating human beings, it is difficult to obtain large enough quantities of interferon to provide effective chemotherapy for viral diseases. (Human donors cannot provide the needed amount of interferon.)

Interferon is a more rapid response to viral infection than antibody response. Interferon production is initiated within hours after viral infection, reaching a maximum after two days. Within three or four days, interferon production declines as antibodies are produced.

Prevention of viral infection by interferon production must be distinguished from another phenomenon—viral interference. Viral interference is observed when an initial viral infection prevents a secondary infection of the same cell by a different virus. The initial virus somehow prevents reproduction of the second virus or inactivates receptors on the cell membrane (there are specific sites for viral attachment). It may also stimulate production of an inhibitor of the second virus. Viral interference does not prevent uninfected cells from becoming infected.

Vaccination involves administration of an attenuated (live, yet weakened) or inactivated strain of virus (or other microorganism), that cannot produce the disease, but that stimulates antibody production and thus prevents infection by the pathogen.

10.7.2 Specific Host Defense Mechanisms—Types of Immunity

Immunity—the ability of the body to recognize and defend itself against an infectious agent.

Specific immunity is characterized by specificity, recognition of self vs. nonself, heterogeneity, and memory.

Heterogeneity is the ability to respond specifically to a variety of substances.

Memory (anamnestic response) is the ability to recognize and respond to a substance previously encountered.

Innate immunity/resistance—genetically predetermined immunity or resistance that an individual is born with, including **species resistance**.

Acquired immunity is specific immunity developed during an individual's lifetime.

Actively acquired immunity involves the production of antibodies or specialized lymphocytes in response to exposure to an antigen; it is usually long lasting.

Passively acquired immunity—antibodies produced by another source are transferred to an individual to confer immunity; they are generally not long lasting.

Naturally acquired active immunity is a result of an infection.

Artificially acquired active immunity is a result of vaccination.

Naturally acquired passive immunity involves transfer of antibodies from mother to fetus (via the placenta) or from mother to newborn (via the colostrum).

Artificially acquired passive immunity involves acquisition of antibodies by injection.

An **antigen** is a chemical substance (usually foreign) that elicits a specific immune response. It may be a protein, glycoprotein, lipoprotein, nucleoprotein, or large polysaccharide.

An **antibody (immunoglobulin)** is a protein produced by B lymphocytes in response to an antigen. Antibodies bind to antigenic determinant sites or epitopes on the antigen. There are five different immunoglobulin (Ig) classes: IgG, IgM, IgA, IgD, and IgE.

IgG antibodies provide naturally acquired passive immunity; they enhance phagocytosis, neutralize toxins and viruses, participate in complement fixation, and protect both fetus and newborn.

IgM antibodies are the first antibodies produced in response to an infection; they are involved in agglutination and complement fixation.

IgA antibodies protect mucosal surfaces.

IgD antibodies appear to be involved in initiation of the immune response.

IgE antibodies are involved in allergic reactions and possibly in responding to protozoal infections.

There are two components of the immune system: **humoral immunity** and **cell-mediated immunity**.

Humoral immunity is involved in defense against toxins, bacteria, and viruses in *extracellular body fluids* such as plasma and lymph.

Cellular immunity (cell-mediated immunity) is involved in the body's response to multicellular parasites, transplanted tissues, cancer cells, and intracellular viruses. T cells do have receptors for antigens, but they do not make antibodies.

Lymphocytes differentiate into either **B cells** (which are involved in humoral immunity) or **T cells** (which are involved in the cell-mediated response).

B-cells are a type of lymphocyte derived from bone marrow stem cells. They later mature to synthesize and secrete antibodies, which are involved in humoral immunity.

T cells are the lymphocytes of the cell-mediated branch of the acquired immune response. T cells mature in the thymus but derive from the progenitor lymphocytes located in the bone marrow. Once T cells have become activated, they differentiate into specialized T cells, each responsible for a different immune response.

Helper T cells are the control cells of the cell-mediated response. These cells contain **CD4 receptors** on the cell surfaces. Their function is to secrete chemical messengers, which regulate the response of other cells in the immune response. Helper T cells are required to activate cytotoxic T cells. B cells must be activated by T helper cells in order to produce antibodies.

Cytotoxic T cells destroy cells that display abnormal cell surface proteins. These cells contain **CD8 receptors** on the cell surfaces. Cytotoxic T cells destroy cells that have been infected with viruses or intracellular bacteria. These cytotoxic T cells also destroy cancer cells. T cells may also be involved in the defense against certain protozoan and helminthic infections.

Delayed hypersensitivity T cells are involved in allergic reactions and in rejection of transplanted tissue. These may also be involved in the defense against cancer.

Suppressor T cells are involved in turning the immune response off. Suppressor T cells are CD8 positive but may not be cytotoxic T cells.

Natural killer cells (**NK cells**) are like T cells in that they are capable of killing other cells. However, unlike T cells, these NK cells are not produced in response to a specific antigen.

Cytokines or **lymphokines** are chemical messengers produced by the cells of the immune system that regulate the immune response. Lymphokines are chemical messengers produced by lymphocytes. The term cytokines refers to all chemical messengers produced by cells. **Interleukins** are chemical messengers that act between leukocytes.

Vaccines and **toxoids** (inactivated toxins) are used to confer active immunization.

Vaccines can be made from live attenuated (weakened) organisms, parts of organisms (subunit vaccines), or dead organisms. **Subunit vaccines** are generally safer than either attenuated organisms or whole killed cells. **Recombinant vaccines**, in which the antigen genes of pathogens are inserted into the DNA of a nonpathogen, are very safe.

Problem Solving Example:

Q A man is exposed to a virus. The virus breaks through the body's first and second lines of defense. Does the body have a third line of defense? Outline the sequence of events leading to the destruction of the virus.

A The first line of defense against invading pathogens, such as viruses, is the skin and mucous membranes. If this line is broken, the second defense mechanism, phagocytosis, takes over. If the virus breaks this line of defense, it enters the circulatory system. Its presence there stimulates specific white blood cells, the lymphocytes, to produce antibodies. These antibodies combine specifically with the virus to prevent it from further spreading.

Any substance, such as a virus, that stimulates antibody synthesis is called an antigen. Antibodies are not produced because the body realizes that the virus will produce a disease, but because the virus is a foreign substance. An individual is "immune" to a virus or any antigen as long as the specific antibody for that antigen is present in the circulatory system. The study of antigen-antibody interactions is called immunology.

The basic sequence of events leading to the destruction of the virus is similar to most antigen-antibody interactions. The virus makes contact with a lymphocyte that has a recognition site specific for the virus. The lymphocyte is then stimulated to reproduce rapidly, causing subsequent increased production of antibodies. The antibodies are released and form insoluble complexes with the virus. These insoluble aggregates are then engulfed and destroyed by macrophages.

10.7.3 Immunological Disorders

Immunological disorders include inappropriate responses (hypersensitivity) and inadequate responses (immunodeficiency).

Hypersensitivity (allergy) is an inappropriate response to an antigen that leads to tissue damage rather than immunity. There are four classes of hypersensitivity: the first three (types I, II, and III) are based on humoral immunity, and the reactions occur within seconds or minutes (type I) or hours (types II and III); type IV, a cell-mediated response, is a delayed reaction, usually occurring within 24–48 hours from the time of exposure.

Type I hypersensitivity (anaphylaxis)—a hypersensitivity reaction involving the production of IgE antibodies that bind to basophils and mast cells; it is characterized by the release of histamine, leukotrienes, and prostaglandins, which cause the characteristic allergic reaction.

Asthma, hives, and hay fever are **localized anaphylactic** reactions; **systemic anaphylaxis**, which may develop rapidly upon antigen exposure, may result in circulatory collapse and death.

Allergies are treated by **hyposensitization (desensitization)**, the repeated injection of increasing concentrations of minute amounts of the allergin. **Antihistamines** are used to treat symptoms.

Type II hypersensitivity (cytotoxic reactions) are the result of mismatched cellular antigens; IgG or IgM antibodies and complement are involved in the destruction of the foreign cells through lysis or phagocytosis.

Transfusion reactions involve mismatches in the ABO blood group or Rh factor and cause cytotoxic reactions.

Hemolytic disease of a newborn occurs when a woman lacking the Rh antigen (said to be Rh-negative) produces anti-Rh antibodies against an Rh-positive fetus.

Type III hypersensitivity (immune complex diseases)—small complexes of IgG antibodies and soluble antigen are deposited in the basement membranes of cells, activating complement. **Complement fixation** then causes inflammation. An **Arthus reaction** is a localized response of this type.

Rheumatoid arthritis, systemic lupus erythematosus, glomerulonephritis, and serum sickness are immune complex disorders.

Type IV hypersensitivity (cell-mediated hypersensitivity) reactions involve T cells, lymphokines, and macrophages, and result in tissue damage.

Contact dermatitis (e.g., from exposure to poison ivy) and tuberculin skin tests are examples of these types of reactions.

Autoimmune disorders involve the development of hypersensitivity to self, as if the self were a foreign substance. They may result in tissue damage from type II, III, or IV reactions. Rheumatoid arthritis and systemic lupus erythematosus are autoimmune diseases.

Histocompatibility antigens occur on the surface membranes of all cells. Human leukocyte antigens (HLAs) are often involved in transplant rejection.

Donor and recipient are matched as closely as possible with regard to HLA and ABO blood group antigens to decrease the chance of rejection.

Patients that have already been sensitized may experience acute graft rejection (a type II response). Slower rejections are usually due to cell-mediated reactions.

Immune deficiencies may be inherited or acquired. There are also many diseases that can impair immune response.

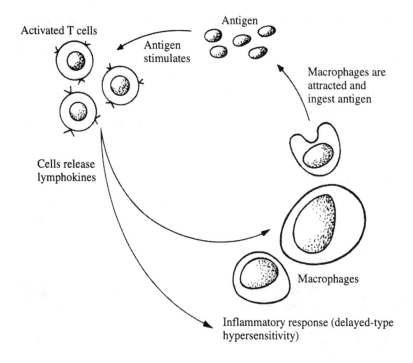

Figure 10.1 Cell-Mediated Immunity

10.7.4 HIV

Human immunodeficiency virus (HIV) is an enveloped retrovirus. An envelope protein is gp120, which attaches to the CD4 cell surface receptor of helper T cells. Once the virus has attached to a CD4 cell, the viral RNA enters the host cell and is transcribed by reverse transcriptase. The resultant DNA molecule is then integrated into the host DNA and begins the production of viral proteins and viral RNAs for repackaging of new virus molecules. Once the virus has reassembled, it is enveloped in the host cell membrane, which contains viral-enveloped proteins, and is once more an infective HIV virus particle.

HIV can be transmitted through infected bodily fluids such as semen, breast milk, and blood. Chemotherapy for HIV infections and treatment of acquired immune deficiency syndrome (AIDS) is directed at reverse transcriptase. Zidovudine (AZT), ddI, and ddC are nucleotide analogues that inhibit reverse transcriptase activity. Because reverse transcriptase is a specific enzyme to the retrovirus, these nucleotide analogues do not affect significantly the normal replication of host DNA. In addition, the protease is a viral protein vital for the production of functional viral proteins. Protease inhibitors stop the action of the protease and prevent the production of these viral proteins.

Researchers are also working to produce an effective vaccine against the HIV virus. The HIV genome is highly susceptible to mutation. This enables the virus to evade chemotherapy treatments and vaccines by constantly varying the chemical constituency of the viral proteins.

Problem Solving Example:

Do people with AIDS make antibodies? If so, how are they immune deficient?

HIV is a retrovirus that attaches to the surface protein CD4. CD4 is a cell surface marker of T helper cells and macrophages. The host produces an active immune response against the invading antigen. Therefore, anti-HIV antibodies are produced by the humoral immune response. However, the T helper cells are compromised because they are the primary target of the virus. T helper cells are responsible for activating the other cells of both the humoral and cell-mediated responses. In Category C of the progression of the HIV infection, T helper cells are reduced to 14% or less of the total T cell population. This results in the immunodeficient state of the host.

10.8 Microbial Diseases of the Skin and Eyes

Bacterial diseases of the skin include staphylococcal, streptococcal, and pseudomonad infections. Most of these organisms are opportunists, generally part of the normal flora, that gain access through cuts, burns, or surgical incisions. Staph infections can be very difficult to treat; at present, there is only one antibiotic (vancomycin) to which *Staphylococcus aureus* is not resistant. In all likelihood, *S. aureus* will develop resistance to vancomycin at some point as well.

Acne is caused by the bacterium *Propionibacterium acnes*, which metabolizes sebum in hair follicles. The fatty acid end products of this metabolism cause inflammation. The condition is treated with tetracycline and benzoyl peroxide.

Viral diseases of the skin include smallpox, chicken pox, shingles, herpes, measles, rubella, and warts.

Fungal skin diseases include athlete's foot, ringworm, and candidiasis.

Diseases of the eye include conjunctivitis, which is caused by a number of bacteria, and keratitis, which can be caused by a herpes simplex type 1 or *Acanthamoeba*.

The eyes of all newborns are treated with a silver nitrate solution to prevent **neonatal gonorrheal ophthalmia**, which is caused by transmission of *Neisseria gonorrhoeae* from an infected mother to her infant during passage through the birth canal.

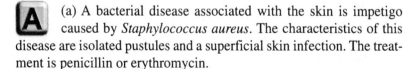

Problem Solving Example:

 Name a bacterial and a viral disease of the (a) skin and (b) eyes.

(a) A bacterial disease associated with the skin is impetigo caused by *Staphylococcus aureus*. The characteristics of this disease are isolated pustules and a superficial skin infection. The treatment is penicillin or erythromycin.

A viral disease associated with the skin is rubella, caused by rubella virus. Rubella or German measles is a mild disease, that resembles measles. There is no treatment.

(b) A bacterial disease of the eye is inclusion conjunctivitis, which is caused by *Chlamydia trachomatis*. The characteristics of this disease are swelling of the eyelid, acute redness and irritation, and associated mucus and pus. The treatment is tetracycline.

A viral disease of the eye is herpetic keratitis caused by herpes simplex type 1 virus. The characteristic is mild conjunctivitis, which may be followed by corneal ulcers. Trifluoridine drops or ointment is an effective treatment.

10.9 Microbial Diseases of the Respiratory System

Infections of the upper respiratory system can be caused by several bacteria and viruses, often in combination. They include pharyngitis, laryngitis, bronchitis, epiglottitis, and sinusitis.

Bacterial diseases of the upper respiratory system include strep throat (caused by group A beta-hemolytic streptococci), scarlet fever, diphtheria, and otitis media (middle ear infections).

The **common cold** is caused by about 200 different viruses; about half are rhinoviruses. Ear infections and sinus infections may occur as complications.

Bacterial diseases of the lower respiratory system include whooping cough, tuberculosis, and pneumonia.

Tuberculosis is a major, worldwide health problem, and its incidence is increasing in the United States. It is caused by *Mycobacterium tuberculosis*.

Pneumonia is often caused by members of the normal flora. Most infections are due to *Streptococcus pneumoniae*, *Haemophilus influenzae*, *Staphylococcus aureus*, *Legionella pneumophila*, and *Mycoplasma pneumoniae*. *Klebsiella* pneumonia, a rarer form of pneumonia, has an 85% mortality rate.

Viral infections cause influenza and pneumonia as well. The influenza virus exhibits antigenic variation.

Fungal diseases of the lower respiratory system, including histoplasmosis, coccidioidomycosis, and blastomycosis, are treated with amphotericin B.

Pneumocystis carinii infects the lower respiratory system of immunosuppressed or immunocompromised individuals. The taxonomic affiliation (fungus or protozoan) is unclear.

Problem Solving Example:

What is Korean hemorrhagic fever and what region in the United States has frequent reports of the fever?

Korean hemorrhagic fever is characterized by the inundation of the lungs by blood plasma and has a five percent mortality rate. It is also the end stage of a severe respiratory disease caused by the *Hantavirus*. The initial symptoms are severe cold or flulike signs. *Hantavirus* is transmitted by inhalation of rodent waste. A majority of human infections have occurred in the Four Corners region of the United States, including several cases reported in 1993 in New Mexico. Intravenous ribavirin is effective in treating severe cases of *Hantavirus* infection. Symptoms are supported as they arise.

10.10 Microbial Diseases of the Digestive System

Tooth decay (dental caries) and **periodontal disease**—*Streptococcus mutans* is involved in the production of plaque. Bacterial acids destroy tooth enamel, and filamentous bacteria and gram-positive rods invade the underlying dentin and pulp. Gram-negative anaerobes, streptococci, and actinomycetes can cause gingivitis and decay of the underlying cementum, leading to periodontal disease.

Gastrointestinal distress can be caused by pathogens growing in the intestine or from the ingestion of toxins (**bacterial intoxication**). Both infections and intoxications can cause diarrhea, dysentery, or gastroenteritis.

Staphylococcal enterotoxicosis—food poisoning due to ingestion of an enterotoxin found in improperly stored foods.

The exotoxin produced by *Staphylococcus aureus* is not denatured by boiling.

Salmonella gastroenteritis is preventable by heating food to 68°C. It can be caused by many *Salmonella* species. Other bacterial diseases of the lower gastrointestinal tract include typhoid fever (due to *Salmonella typhi* infection), cholera (due to *Vibrio cholerae* exotoxin), *Vibrio parahaemolyticus* gastroenteritis from contaminated mollusks or crustaceans, and *Escherichia coli* gastroenteritis (traveler's diarrhea).

Helicobacter pylori is the cause of **peptic ulcer disease (PUD)**. *H. pylori* binds to the epithelium of the stomach and can grow in the highly acidic environment. This infection causes an immune response and inflammation. In combination with the acid content of the stomach, the inflammation can progress to ulcerated tissue and PUD. The treatment of PUD is now antibiotic treatment in combination with bismuth subsalicylate.

Viral diseases of the digestive tract include mumps and hepatitis.

Fungal diseases of the digestive system include ergot poisoning and aflatoxin poisoning. Mycotoxins (fungal toxins) can affect the blood, nervous system, liver, or kidneys.

There are a number of protozoan diseases of the digestive system, including *Giardia lamblia* infection (giardiasis), amoebic dysentery (amoebiasis), and *Cryptosporidium* infection.

Helminths also cause diseases in the gastrointestinal tract: nematode infestations (pinworms, hookworms, and trichinosis) and tapeworm infestation.

Problem Solving Example:

 A microbiologist takes a stool specimen from you and a few days later tells you that *Escherichia coli* and *Salmonella typhi* are growing in your intestinal tract. Which type should you be concerned about?

A *Escherichia coli,* or *E. coli,* is a species of bacteria that normally inhabits the intestine of humans and other animals; therefore, the presence of *E. coli* in the stool should not cause much alarm. In adults, the infections caused by *E. coli* are usually not severe. For example, they may cause diarrhea, and are occasionally found in infections of the urogenital tract. However, in infants, children, and the elderly, the infections can require hospitalization and in some cases be life threatening.

The genus *Salmonella,* however, includes several species that are pathogenic to man and other animals. *Salmonella typhi,* for example, causes the acute infectious disease known as typhoid fever. Typhoid fever is characterized by a fever, inflammation of the intestine, intestinal ulcers, and a toxemia (presence of toxins in the blood). The presence of *Salmonella* is therefore ample reason for concern.

Typhoid fever, like most intestinal infections, is transmitted from one person to another through food and water. Transmission may be indirect. For example, wastes from an infected person can pollute drinking water or food, or the infected person may handle food at some point in its processing or distribution and contaminate it, affecting the consumer. The common housefly can also transmit *Salmonella* from wastes to food. Typhoid fever occurs in all parts of the world. In locations where good sanitation (proper disposal and treatment of biological waste and water purification) is practiced, typhoid fever incidence is very low. Carriers are people who are infected with *Salmonella typhi,* but who have had only a slight intestinal infection, and hence do not know they harbor the pathogenic organism. Carriers should not be allowed to handle or prepare food.

10.11 Microbial Diseases of the Cardiovascular System

Septicemia—growth of microorganisms in the blood—can cause inflamed lymph vessels (**lymphangitis**), septic shock, and decreased blood pressure. Symptoms are usually the result of endotoxins.

Puerperal sepsis is a uterine infection following childbirth or abortion; it is usually caused by *Streptococcus pyogenes*.

Bacterial endocarditis—infection (usually streptococcal or staphylococcal) of the inner layer of the heart.

Rheumatic fever—a possible complication following streptococcal infection.

Bacteria, such as *Pasteurella multocida*, can be introduced by animal bites and scratches (especially cats, dogs, and rats).

Viral diseases include infectious mononucleosis, yellow fever, and dengue virus.

Protozoans cause toxoplasmosis, malaria, and Chagas' disease.

Schistosomiasis and swimmer's itch are helminthic diseases.

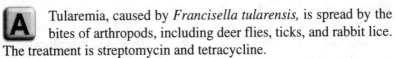

Problem Solving Example:

List four diseases that are transmitted by animal bites.

Tularemia, caused by *Francisella tularensis,* is spread by the bites of arthropods, including deer flies, ticks, and rabbit lice. The treatment is streptomycin and tetracycline.

Rat-bite fever, caused by *Streptobacillus* or *Spirillum,* causes a disease whose signs and symptoms include fever, inflammation, pain and stiffness of the joints, and infections of the lymphatic system. The treatment is penicillin or tetracycline.

Cat scratch disease may be caused by *Afipia felis* or *Bartonella henselae,* and is spread by cat scratches or bites. The characteristics include rash, fever, and swelling of the lymph nodes. Rifampin is used to treat this disease.

Lyme disease is caused by the spirochete *Borrelia burgdorferi* and is spread by ticks. The treatment is penicillin and tetracycline.

10.12 Microbial Diseases of the Nervous System

Bacterial meningitis can be caused by nearly 50 species of opportunistic bacteria, including *Neisseria meningitidis*, *Haemophilus influenzae*, and *Streptococcus pneumoniae*.

Leprosy, tetanus, and botulism are also caused by bacteria.

Viral diseases include poliomyelitis, rabies, and arthropod-borne encephalitis.

Fungal diseases include *Cryptococcus* meningitis.

Protozoal diseases include *Naegleria* meningoencephalitis and African trypanosomiasis.

10.13 Microbial Diseases of the Genitourinary System

Bacterial diseases of the urinary system include cystitis, glomerulonephritis, and pyelonephritis.

Bacterial diseases of the reproductive system include gonorrhea, nongonococcal urethritis, syphilis, vaginitis, chancroid, granuloma inguinale, and lymphogranuloma venereum.

Viral diseases of the reproductive system include cytomegalovirus, genital herpes, and genital warts.

Candidiasis, a fungal disease, and trichomoniasis, a protozoal disease, also affect the genitourinary system.

Problem Solving Example:

 A newborn infant is examined by a doctor, who determines that the child has syphilis. Explain how the child contacted syphilis.

Syphilis is a venereal disease of humans that is included in the more general category of contact diseases. Contact diseases of humans usually result from the entry of an infectious agent into the individual through the skin or mucous membranes. Contact may be direct (through wounds and abrasions) or indirect (through vectors— mediating transmitters such as insects). In the case of venereal diseases, transmittance is usually through direct genital contact during sexual activity. The organism involved is a spirillum called *Treponema pallidum.*

However, direct sexual contact is not the only way in which syphilis is transmitted. An infected mother can transmit the organism by placental transfer to the fetus during the first four months of pregnancy. Contraction of the disease is also possible as the fetus passes through the infected vagina during birth.

The disease usually requires an incubation period of three to six weeks after infection. The disease produces lesions called chancres that resemble ulcerous sores. In the late stages of the disease, the cardiovascular and central nervous systems may be affected, with possible paralysis. The disease, if promptly detected, can be treated with penicillin. There is no known means for immunization against *T. pallidum* infection. Persons who have recovered from a syphilitic infection are just as likely to contract it upon subsequent exposure to the organism. Preventive measures include avoiding carriers and using local prophylactic measures such as condoms.

Quiz: Role of Microbes in Disease

1. Photosynthetic algae and water-capturing fungi combine to form lichens. The algae provide nutrients and the fungi supply water. This is an example of

    (A) parasitism.

    (B) mutualism.

    (C) predation.

    (D) commensalism.

    (E) competition.

2. Pathogenic bacteria

    (A) are autotrophs.

    (B) are beneficial to humans.

    (C) cause disease.

    (D) die easily.

    (E) live constantly in the soil.

3. A microorganism suspected of causing a disease in a host is discovered. It is isolated and grown in culture. Microorganisms from the culture are then injected into a healthy host. What must be done in order to determine whether or not the microorganism is a pathogen?

    (A) The microorganism's karyotype must be determined.

    (B) The toxin must be extracted from the microorganism.

    (C) The microorganism must be isolated and grown in culture; the new culture must then be compared to the original culture.

(D) A Gram stain and motility test must be performed on the organism.

   (E) The size, shape, motility, and nutritional requirement of the organism must be determined.

4.   T cells are generally NOT involved in fighting

   (A) cancer cells.

   (B) transplanted foreign tissue.

   (C) viral infections.

   (D) bacterial infections.

   (E) parasitic infections.

5.   A hookworm bores through the skin of a human, migrates to the intestine, attaches, and feeds on blood. The hookworm is a

   (A) commensal.

   (B) parasite.

   (C) predator.

   (D) saprophyte.

   (E) scavenger.

6.   A protein produced by virus-infected vertebrate cells that spreads to neighboring cells and helps protect them from the virus is a(n)

   (A) antibiotic.

   (B) antibody.

   (C) antigen.

   (D) complement.

   (E) interferon.

7. The procedures used by Robert Koch were used to discover the cause of which of the following diseases in humans?

    (A) Bubonic plague

    (B) Cholera

    (C) Tuberculosis

    (D) Leprosy

    (E) Tetanus

8. A certain bacterial infection is characterized by extreme virulence due to toxins released by bacteria. The toxin cannot be inactivated by heating or by decreasing the pH. It may be concluded that

    (A) the bacterium is probably gram-positive and is elaborating an exotoxin.

    (B) the bacterium is probably gram-negative and is producing an endotoxin.

    (C) the bacterium is probably gram-positive; the virulence is due to an endotoxin.

    (D) the symptoms are caused by a gram-negative bacterium and its exotoxin.

    (E) None of the above.

9. When certain bacteria are removed from the human intestinal tract, which one of the following cannot be synthesized?

    (A) Riboflavin

    (B) Vitamin B_{12}

    (C) Thiamin

    (D) Glucosamine

    (E) Myosin

10. A worm called *Schistosoma mansoni* is found in the Nile Valley of Egypt. The extensive irrigation canals fed by the Aswan Dam have allowed aquatic snails to flourish; these snails serve as intermediate hosts to the worm. *Schistosoma mansoni* enters the human body right through the skin when people wade in the water. The worm then feeds on the nutrients in the human blood. The worm's relationship to humans is

   (A) competition.

   (B) mimicry.

   (C) mutualism.

   (D) parasitism.

   (E) allelopathy.

ANSWER KEY

1.	(B)		6.	(E)
2.	(C)		7.	(B)
3.	(C)		8.	(B)
4.	(D)		9.	(B)
5.	(B)		10.	(D)

CHAPTER 11

Microbes in the Environment

11.1 Microbes and the Recycling of Nutrients

Microbes, especially bacteria and fungi, play an important role in the **decomposition** of organic matter and the **recycling** of chemical elements.

Bioconversion is the microbial conversion of organic waste materials into alternative fuels.

Bioremediation is the use of bacteria to clean up toxic wastes.

11.1.1 Biogeochemical Cycles

Microbes are essential to the **recycling** of chemical elements such as carbon, nitrogen, phosphorus, sulfur, and oxygen.

They also **solubilize minerals**, such as sulfur, potassium, iron, and others, making them available for plant metabolism.

The carbon cycle—photoautotrophs fix CO_2, providing nutrients for chemoheterotrophs. Chemoheterotrophs release CO_2, which can then be used by the photoautotrophs.

The nitrogen cycle—bacteria are involved in the decomposition of proteins from dead cells, ammonification of the amino acids, and reduction of nitrates to molecular nitrogen (N_2). Nitrogen-fixing bacteria are responsible for converting molecular nitrogen back into ammonium and nitrate, which can then be used by other bacteria and plants in the synthesis of amino acids.

Problem Solving Example:

Discuss the role of bacteria in the carbon cycle and the nitrogen cycle.

Cycling of the earth's resources is a process by which life is able to continue on earth. The carbon and nitrogen atoms that are present on the earth today are the same atoms that were present three billion years ago, and have been used over and over again. They are the fundamental constituents of organic compounds, and are present in large quantities in all organisms. The same carbon and nitrogen atoms are constantly being recycled.

In the carbon cycle, atmospheric carbon dioxide is converted into organic material by plants. Animals and bacteria convert some of this organic matter into CO_2 by respiration; however, most of the carbon remains fixed as organic matter in the bodies of plants and animals. Bacteria play a crucial role in the carbon cycle. Bacteria, as well as fungi, convert the carbon atoms in the organic matter of decaying plants, animals, and other bacteria and fungi to CO_2, which returns to the atmosphere.

Plants obtain nitrates from the soil, taking them up through their roots, and convert the nitrate into organic compounds, mainly proteins. Animals obtain nitrogen from plant proteins and amino acids, and they excrete nitrogen-containing wastes. These nitrogenous wastes are excreted in one of the following forms, depending on the species—urea, uric acid, creatinine, and ammonia. Certain bacteria in the soil convert nitrogenous waste and the proteins of dead plants and animals into ammonia. Another type of bacteria present in the soil is able to convert ammonia to nitrate. These are termed nitrifying bacteria. They obtain energy from chemical oxidations. There are two types of nitrifying bacteria: nitrite bacteria, which convert ammonia into nitrite, and nitrate bacteria, which convert nitrite into nitrate. Nitrogen is thus returned to the cycle. Atmospheric nitrogen, N_2, cannot be utilized as a nitrogen source by either animals or plants. Only some blue-green algae and certain bacteria can convert N_2 to organic compounds. This process is termed nitrogen fixation. One genus of bacteria, *Rhizobium*, is able to utilize N_2 only when grown in association with leguminous plants, such as peas and beans. The bacteria grow inside tiny swellings of the plant's roots, called root nodules. Nitrogen is also returned to the atmosphere by certain bacteria. Denitrifying bacteria convert nitrites and nitrates to N_2, thus preventing animals and plants from obtaining biologically useful nitrogen.

11.1.2 Microbes in the Soil, Water, and Air

Microbes from all taxonomic groups are present in the soil.

Soil contains inorganic material (rocks, minerals, water, and gases) as well as organic matter (humus) and microorganisms.

Microbes from all taxonomic groups are present in both freshwater and marine environments. However, the high osmotic pressure, low nutrient availability, and high pH of the open ocean makes the marine environment unfavorable for many microbes.

Many pathogens are transmitted in drinking water. Bacterial counts are used in assaying water purity.

Sewage treatment—aerobic bacteria are used to decompose organic matter in secondary treatment of sewage waste.

Microbes do not live in the air, but can be transmitted through the air.

Problem Solving Example:

 List some of the soil microbiota, microbial pathogens in soil, freshwater, and seawater microbiota.

The most abundant microbiota found in soil is bacteria, including actinomycetes. Fungi are also found in soil—molds more than yeast. There are much smaller numbers of fungi than bacteria found in soil; however, the fungal biomass is about equivalent to the bacterial biomass. This is due to the larger size of the fungi. Protozoa and algae are also abundant in soil.

Endospore-forming bacteria constitute a large number of the microbial pathogens found in soil, including *Bacillus anthracis* (anthrax), *Clostridium tetani* (tetanus), *Clostridium botulinum* (botulism), and *Clostridium perfringens* (gas gangrene). Helminths may spend part of their life cycle in soil. The larval stage of the hookworm lives in soil.

Plant and insect pathogens also inhabit soil.

Freshwater microbiota include photosynthetic algae, pseudomonads, *Cytophaga, Caulobacter,* and *Hyphomicrobium* in the limnetic zone. Purple and green sulfur bacteria are found in the profundal zone. *Desulfovibrio* is found in the benthic mud, while methane-producing bacteria are found in the benthic zone.

Phytoplankton, algae, and some bacteria are a part of the seawater microbiota. The conditions of the ocean, high osmotic pressure, low nutrients, and high pH are not favored by many microorganisms.

11.2 Bioremediation

Bioremediation is the concerted effort by researchers and scientists to increase the activity of those microorganisms that effectively degrade pollution. Certain types of bacteria are able to degrade petroleum but require the addition of nitrogen and phosphorus to encourage bacterial growth. Oxygen is also required for this process, and contaminated soil must be constantly aerated in order to encourage bacterial degradation of petroleum spills.

Bacteria also are able to naturally degrade or chemically process heavy metals, sulfur, nitrogen gas, and polychlorinated biphenyls (PCBs). These bacteria may also be genetically altered to increase their natural ability to degrade these pollutants. This solution to pollution is attractive because it converts a potentially harmful substance to a harmless or useful substance.

Pseudomonads are capable of converting methyl mercury, a highly toxic form of mercury, to mercuric ion and then to elemental mercury, a relatively nontoxic element.

Problem Solving Example:

 Explain why nitrogen and phosphorus are added to beaches following an oil spill.

 Nitrogen and phosphorus are used to encourage the growth of natural oil-degrading bacteria. Nitrogen is about 12–15% of the dry weight of a bacterial cell, and phosphorus and sulfur make up about 3%. The synthesis of nucleic acids and amino acids requires nitrogen. Sulfur is needed for the synthesis of some amino acids and vitamins. Bacteria can extract pollutants that have combined with soil and water and may change the chemistry and, therefore, the toxicity of the pollutant. *Pseudomonas* are naturally occurring microorganisms that can degrade oil.

CHAPTER 12

Microbes in Industry

12.1 Microbes in the Food Industry

Alcoholic beverages and vinegar—fermentation by yeasts is responsible for the production of ethanol; *Acetobacter* and *Gluconobacter* oxidize the alcohol in wine to acetic acid (vinegar).

Cheese—Lactic acid bacteria are used to curdle cheese and in the production of hard cheeses.

Lactobacilli, streptococci, and yeasts are used in the production of buttermilk, sour cream, and yogurt.

Nondairy fermentation by various microbes is involved in making sauerkraut, pickles, olives, and soy sauce. The fermentation of yeast produces ethanol and CO_2, which makes bread dough rise.

Problem Solving Example:

In what ways are bacteria useful to the dairy industry?

Bacteria are a major cause of disease, but beneficial bacterial species outnumber harmful ones. There are many useful applications of microorganisms in the dairy industry. Fermented milk products are made by encouraging the growth of the normal lactic acid-producing bacteria that are present in the culture. Specific organisms, or mixtures of them, are used to produce buttermilk, yogurt, and other fermented milk products. The principal bacteria used are species of *Streptococcus, Leuconostoc,* and *Lactobacillus,* which produce only lactic acid. Butter is made by churning pasteurized sweet or sour cream, with the latter being fermented by streptococci and leukonostocs. Different types of cheeses are made by providing conditions that favor the development of selected types of bacteria and molds. The quality and characteristics of a cheese are determined by the biochemical activities of selected microorganisms. For example, Swiss cheese is produced through the fermentation of lactic acid by bacteria of the genus *Propionibacterium.* The products of the fermentation are propionic acid, acetic acid, and carbon dioxide. The acids give the characteristic flavor and aroma to the cheese, and the accumulation of carbon dioxide produces the familiar holes.

Bacteria are also useful to the dairy industry by providing the means by which certain milk-giving animals acquire nutrients. The ruminants (cattle, sheep, goats, and camels) are a group of herbivorous mammals whose digestive system has a compartmentalized stomach, the first section being termed the rumen. In the rumen, there exist certain bacteria and protozoa, which provide enzymes necessary to break down the cellulose acquired from the ingestion of plant material. The mammal does not normally have the enzymes needed to degrade cellulose into nutrients that the animal can use (the lack of these enzymes in humans also explains why humans cannot break down plant material).

Bacteria are also important to the food industry in other ways. Pickles, sauerkraut, and some sausages are produced in whole or part by microbial fermentations.

12.2 Industrial Microbiology

Industrial microbiology—use of microbes to manufacture or help manufacture useful products or to dispose of waste.

Microbes can produce acetone, alcohols, glycerol, and organic acids.

Some microbes, e.g., *Thiobacillus ferrooxidans*, can extract mineral ores. Minerals that can be extracted include arsenic, copper, uranium, iron, cobalt, lead, nickel, and zinc.

A large number of enzymes are commercially produced by microorganisms. For example, **alpha amylase** is produced by *Aspergillus niger* or *Aspergillus oryzae*. This enzyme is used as a flour supplement in baking, in the textile industry, and in the pharmaceutical industry as a digestive aid for beans. *Bacillus subtilis* is a major producer of alpha amylase used in brewing. The enzyme **beta amylase** is produced by a variety of microorganisms and is used for the production of maltose syrup. **Invertase** is produced by *Saccharomyces cervisiae* in the production of candy and artificial honey. **Streptokinase** is produced by *Streptococcus* (spp.) as an exotoxin and is used in medicine to lyse embolisms and thrombolisms.

Problem Solving Example:

What is the role of microorganisms in the production of industrial chemicals and pharmaceuticals?

Corynebacterium glutamicum produces lysine and glutamic acid. *Aspergillus niger* produces citric acid. Alpha amylase produced by *Bacillus* bacteria as well as the fungi *Aspergillus niger* and *Aspergillus oryzae,* is used as a flour supplement, in the breakdown of starch, in the textile industry, and also in the pharmaceutical industry to assist in the digestion of beans. *Pseudomonas* and *Propionibacterium* produce vitamin B_{12}. Riboflavin is produced by *Ashbya gossypii*. The pharmaceutical industry uses microorganisms to produce antibiotics such as penicillin and vaccines for use in medicine. *Streptomyces,* a genus of aerobic, moldlike bacteria, produce steroids. *Thiobacillus ferrooxidans* is used in the recovery of poor grades of uranium and copper.

12.3 Microbes and Medicine

Pharmaceutical microbiology—use of microbes to manufacture products used in medicine.

Bacteria produce most of the amino acids used in medicine and food.

Microbes can produce antibiotics, enzymes, vitamins, and hormones.

Problem Solving Example:

 What is reverse transcriptase, and what is its function? How might it be used in cancer research?

We have seen in protein synthesis that biological information flows from DNA to RNA to protein. This flow had become the rule until 1964, when Temin found that infection with certain RNA tumor viruses (cancer-causing substances) is blocked by inhibitors of DNA synthesis and by inhibitors of DNA transcription. This suggested that DNA synthesis and transcription are required for the multiplication of RNA tumor viruses. This would mean that the information carried by the RNA of the virus is first transferred to DNA, whereupon it is transcribed and translated, and consequently, that information flows in the reverse direction, that is, from RNA to DNA. Temin proposes that the RNA of these tumor viruses, in their replication, are able to form DNA. His hypothesis requires a new kind of enzyme—one that would catalyze the synthesis of DNA using RNA as a template. Such an enzyme was discovered by Temin and by Baltimore in 1970. This RNA-directed DNA polymerase, also known as reverse transcriptase, has been found to be present in all RNA tumor viruses.

An infecting RNA virus binds to and enters the host cell (the cell that the virus attacks). Once in the cytoplasm, the RNA genome is separated from its protein coat. Then, through an as yet unknown mechanism, the viral reverse transcriptase is used to form a DNA molecule using the viral RNA as a template. This DNA molecule is integrated into the host's chromosome, with a number of possible consequences. The viral DNA may now be duplicated along with the host's DNA, and its information thus propagated to all offspring of the infected cell. Its presence in the genome of the host may "transform" the cell and its offspring (causing them to become cancerous). In addition, the viral

DNA may be used as a template for the synthesis of new viral RNA, and the virus thus multiplies itself and continues its infectious process, often without killing the cell. Because the viral genome is RNA, it cannot as such be integrated into the host's genome. Reverse transcriptase enables the virus to convert its genetic material to DNA, whereupon it is capable of inserting itself into the host chromosome.

There are three reasons why reverse transcriptase is so important in cancer research. First, human leukemia sarcomas (different forms of cancer) have been shown to contain large RNA molecules similar to those of the tumor viruses that cause cancer in mice. Second, these human cancer cells contain particles with reverse transcriptase activity. Third, the DNA of some human cancer cells have viruslike sequences of nucleotides, sequences not found in the DNA of comparable normal cells.

Since the action of RNA from tumor viruses depends on the presence of the enzyme reverse transcriptase, research into its chemical composition (amino acid sequence) may bring better understanding of its function. The ultimate objective of such research would of course be the prevention of cancer, perhaps through the inhibition of reverse transcriptase activity.

12.4 Microbes and Recombinant DNA Technology

The ability to genetically engineer cells has paved the way for the production of many new products.

12.4.1 Applications in Medicine

Currently, such important substances as insulin and interferon are produced from genetically engineered microorganisms.

Synthetic genes for the two polypeptides that make up **human insulin** have been inserted into a plasmid vector and transferred into *E. coli* for production and secretion. Mammalian cells are also used to produce insulin.

Cells can be engineered to produce surface proteins of pathogens for use as **subunit vaccines**. Animal viruses can be used to engineer **recombinant vaccines**, e.g., the antigen genes from a pathogenic microorganism can be inserted into the DNA of a nonpathogenic microbe.

The **polymerase chain reaction** is used to enzymatically produce multiple copies of a piece of DNA, and to increase amounts of DNA in samples to detectable levels so that they may be used in gene sequencing and other diagnostic procedures.

DNA probes provide rapid identification of pathogens in food and body tissues.

Problem Solving Example:

 Why is the organism that causes smallpox not used in the vaccine against smallpox?

Smallpox (variola) is an acute infectious disease that is caused by a virus. The disease is spread by droplet infection or by handling articles infected by a smallpox patient. The virus is thought to lodge in the nasopharynx (the part of the pharynx just dorsal to the soft palate), where it proliferates and spreads to the blood (viremia), enabling it to infect the skin and other tissues. The disease is characterized by an initial fever followed by the appearance of pustules (small fluid-filled eruptions) on the skin, which regress and leave the scars characteristic of smallpox.

It has long been known that contracting smallpox protects one against a second attack. For centuries, the Chinese inoculated the skin of a healthy person with the material from lesions on a smallpox patient. The variola virus, present in this material, would stimulate antibody production in the inoculated individual to give the person immunity to smallpox. However, the use of the actual virulent variola virus may be dangerous if too much is administered. In 1796, it was discovered by Jenner that inoculation with the vaccinia virus from cowpox lesions would confer immunity to smallpox as well as to cowpox. This immunization procedure is safe because smallpox is never produced.

This vaccination procedure is practiced today, using vaccinia virus taken from cows, sheep, or chick embryos.

Although the United States and many other countries have eliminated smallpox, outbreaks can occur as long as reservoirs of the variola virus still exist (e.g., Africa, India, and Southeast Asia). However, today the smallpox virus is basically confined to research laboratories. People are usually vaccinated very early in life then revaccinated every three to five years. Prevention is vital since there is no specific treatment for smallpox.

12.4.2 Applications in Agriculture

Cells from plants with the desired characteristics can be cloned.

The vector most often used for the transfer of plant DNA is the **Ti plasmid** inserted into *Agrobacterium*. *Agrobacterium* is a bacterium that naturally transforms plant cells. Thus, *Agrobacterium* readily transfers genes to plants and, as such, is a very useful tool in genetic engineering.

E. coli is used to produce bovine growth hormone.

Problem Solving Example:

 List some other agriculturally important applications of genetic engineering.

 Pseudomonas syringae protects against the frost damage of plants because it cannot produce a protein that induces ice-crystal formation.

Pseudomonas fluorescens has been engineered to contain an insect pathogen from *Bacillus thuringiensis*. This pathogen kills insects that eat the roots of plants.

Rhizobium meliloti has been engineered to enhance nitrogen fixation.

12.4.3 Safety Issues

There are strict safety standards for the development and use of genetically engineered microorganisms.

One technique for decreasing the probability that a genetically engineered microbe will survive outside the lab or beyond a certain time limit is to insert so-called **suicide genes** into them along with the gene of interest.

12.4.4 The Future of Genetic Engineering

Genetic engineering should continue to provide us with new diagnostic tools and treatments for disease.

Quiz: Microbes in the Environment and Industry

1. All of the following processes occur in the nitrogen cycle EXCEPT

    (A) ammonification.

    (B) nitrification.

    (C) deamination.

    (D) denitrification.

2.   What group of bacteria belongs in circle 4?

    (A) Nitrogen-fixing bacteria

    (B) Plankton

    (C) Nitrifying bacteria

    (D) Denitrifying bacteria

    (E) Decomposing bacteria

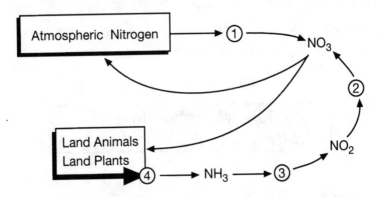

3.   The conversion of atmospheric nitrogen by certain eubacteria and blue-green bacteria into a form usable by other living organisms is termed

    (A) ammonification.

    (B) denitrification.

    (C) nitrification.

    (D) nitrogen fixation.

    (E) transformation.

4.   A protein currently synthesized by bacterial cells that have been altered by gene splicing is

    (A) actin.

    (B) AMP.

    (C) hemoglobin.

    (D) insulin.

    (E) myosin.

5.   Cloning is a reproductive process which involves (a)

    (A) sexual reproductive process.

    (B) formation of genetic duplicates.

    (C) combination of meiosis and mitosis.

    (D) production of genetic variability.

    (E) yielding sex cells.

6.   Lambda phages are being used as vectors to deliver recombinant DNA into bacteria for the purpose of DNA cloning. Which of the following is or are the advantage(s) of choosing lambda as a vector?

    (A) There is no limitation on the length of foreign DNA being inserted into the lambda genome.

    (B) Lambda phage infects bacteria with a high frequency.

    (C) Large segments of the lambda genome are not essential for its lytic and lysogenic life cycles.

    (D) Both (A) and (B).

    (E) Both (B) and (C).

7.   Fecal coliforms are indicators of water contamination by

    (A) enteric bacteria.

    (B) enteric viruses.

    (C) enteric protozoa.

    (D) enteric helminths.

    (E) None of the above.

8. Which of the following is a potential use of genetic engineering in agriculture?

   (A) The production of bovine growth factor to increase the weight of and milk production in cattle.

   (B) Use of microorganisms (capable of degrading airborne pollutants) in soil of potted plants.

   (C) Protection of strawberries by spraying them with an ice resistant bacterium.

   (D) Both (A) and (B).

   (E) All of the above.

9. Which is a tumor-producing bacterium that infects plants and may be of use in genetic engineering in agriculture?

   (A) Endomycorrhizae

   (B) *Aspergillus niger*

   (C) *Pseudomonas fluorescens*

   (D) Ti plasmid

   (E) *Agrobacterium*

10. Nitrifying bacteria have been isolated from

   (A) sewage disposal systems.

   (B) freshwater habitats.

   (C) seawater habitats.

   (D) soil.

   (E) All of the above.

ANSWER KEY

1.	(C)	6.	(E)
2.	(E)	7.	(A)
3.	(D)	8.	(E)
4.	(D)	9.	(E)
5.	(B)	10.	(E)

THE BEST TEST PREPARATION FOR THE

AP*
ADVANCED
PLACEMENT
EXAMINATION

BIOLOGY

6 Full-Length Practice Exams

Based on official exams released by the College Board

Detailed explanations to every exam question

Far more comprehensive than any other test preparation book

Includes a **COMPREHENSIVE REVIEW COURSE** of the topics covered on the exam. This book can be used for self-study or by any class preparing for the exam.

REA *Research & Education Association*

* AP is a registered trademark of the
College Entrance Examination Board,
which does not endorse this book.

Available at your local bookstore or order directly from us by sending in coupon below.

REA's Test Preps
The Best in Test Preparation

- REA "Test Preps" are **far more** comprehensive than any other test preparation series
- Each book contains up to **eight** full-length practice tests based on the most recent exams
- **Every** type of question likely to be given on the exams is included
- Answers are accompanied by **full** and **detailed** explanations

REA publishes over 60 Test Preparation volumes in several series. They include:

Advanced Placement Exams(APs)
Biology
Calculus AB & Calculus BC
Chemistry
Computer Science
Economics
English Language & Composition
English Literature & Composition
European History
Government & Politics
Physics B & C
Psychology
Spanish Language
Statistics
United States History

College-Level Examination Program (CLEP)
Analyzing and Interpreting Literature
College Algebra
Freshman College Composition
General Examinations
General Examinations Review
History of the United States I
History of the United States II
Human Growth and Development
Introductory Sociology
Principles of Marketing
Spanish

SAT II: Subject Tests
Biology E/M
Chemistry
English Language Proficiency Test
French
German

SAT II: Subject Tests (cont'd)
Literature
Mathematics Level IC, IIC
Physics
Spanish
United States History
Writing

Graduate Record Exams (GREs)
Biology
Chemistry
Computer Science
General
Literature in English
Mathematics
Physics
Psychology

ACT - ACT Assessment

ASVAB - Armed Services Vocational Aptitude Battery

CBEST - California Basic Educational Skills Test

CDL - Commercial Driver License Exam

CLAST - College Level Academic Skills Test

COOP & HSPT - Catholic High School Admission Tests

ELM - California State University Entry Level Mathematics Exam

FE (EIT) - Fundamentals of Engineering Exams - For both AM & PM Exams

FTCE - Florida Teacher Certification Exam

GED - High School Equivalency Diploma Exam (U.S. & Canadian editions)

GMAT CAT - Graduate Management Admission Test

LSAT - Law School Admission Test

MAT - Miller Analogies Test

MCAT - Medical College Admission Test

MTEL - Massachusetts Tests for Educator Licensure

MSAT - Multiple Subjects Assessment for Teachers

NJ HSPA - New Jersey High School Proficiency Assessment

NYSTCE: LAST & ATS-W - New York State Teacher Certification

PLT - Principles of Learning & Teaching Tests

PPST - Pre-Professional Skills Tests

PSAT - Preliminary Scholastic Assessment Test

SAT I - Reasoning Test

TExES - Texas Examinations of Educator Standards

THEA - Texas Higher Education Assessment

TOEFL - Test of English as a Foreign Language

TOEIC - Test of English for International Communication

USMLE Steps 1,2,3 - U.S. Medical Licensing Exams

U.S. Postal Exams 460 & 470

RESEARCH & EDUCATION ASSOCIATION
61 Ethel Road W. • Piscataway, New Jersey 08854
Phone: (732) 819-8880   website: www.rea.com

Please send me more information about your Test Prep books

Name _____

Address _____

City _____ State _____ Zip _____